NOV 14

THE WORLD BOOK ENCYCLOPEDIA OF
PEOPLE AND PLACES

6
T-Z

Prospect Heights Public Library
12 N Elm Street
Prospect Heights, IL 60070
www.phpl.info

a Scott Fetzer company
Chicago

www.worldbookonline.com

For information about other World Book publications,
visit our website at http://www.worldbookonline.com
or call 1-800-WORLDBK (1-800-967-5325).

For information about sales to schools and libraries, call
1-800-975-3250 (United States);
1-800-837-5365 (Canada).

The World Book Encyclopedia of People and Places,
second edition
© 2013 World Book, Inc. All rights reserved. This volume may not be reproduced in whole or in part in any form without prior written permission from the publisher.

WORLD BOOK and the GLOBE DEVICE are registered trademarks or trademarks of World Book, Inc.

Copyright © 2011, 2008, 2007, 2005 World Book, Inc.;
© 2004, 2003, 2002, 2000, 1998, 1996, 1995, 1994,
1993, 1992 World Book, Inc., and Bertelsmann GmbH.
All rights reserved.

Library of Congress Cataloging-in-Publication Data

The World Book encyclopedia of people and places.
v. cm.
Summary: "A 7-volume illustrated, alphabetically arranged set that presents profiles of individual nations and other political/geographical units, including an overview of history, geography, economy, people, culture, and government of each. Includes a history of the settlement of each world region based on archaeological findings; a cumulative index; and Web resources"--Provided by publisher.
Includes index.
ISBN 978-0-7166-3758-5
1. Encyclopedias and dictionaries. 2. Geography--Encyclopedias. I. World Book, Inc. Title: Encyclopedia of people and places.
AE5.W563 2011
030--dc22
2010011919

This edition ISBN: 978-0-7166-3760-8

Printed in Hong Kong by Toppan Printing Co. (H.K.) LTD
3rd printing, revised, August 2012

Cover image:
Great Plains, United States

© Peter Beck, Corbis

STAFF

Executive Committee:
President: Donald D. Keller
Vice President and Editor in Chief: Paul A. Kobasa
Vice President, Marketing & Digital Products: Sean Klunder
Vice President, International: Richard Flower
Controller: Yan Chen
Director, Human Resources: Bev Ecker

Editorial:
Associate Director, Supplementary Publications: Scott Thomas
Senior Editors: Lynn Durbin, Kristina Vaicikonis, Daniel O. Zeff
Senior Statistics Editor: William M. Harrod
Editor: Mellonee Carrigan
Researcher, Supplementary Publications: Annie Brodsky
Manager, Indexing Services: David Pofelski
Manager, Contracts & Compliance (Rights & Permissions): Loranne K. Shields
Administrative Assistant: Ethel Matthews

Editorial Administration:
Director, Systems and Projects: Tony Tills
Senior Manager, Publishing Operations: Timothy Falk
Associate Manager, Publishing Operations: Audrey Casey

Manufacturing/Production:
Director: Carma Fazio
Manufacturing Manager: Steven K. Hueppchen
Production/Technology Manager: Anne Fritzinger
Senior Production Manager: Jan Rossing
Production Specialist: Curley Hunter
Proofreader: Emilie Schrage

Graphics and Design:
Senior Manager: Tom Evans
Coordinator, Design Development and Production: Brenda B. Tropinski
Senior Designer: Isaiah W. Sheppard, Jr.
Associate Designer: Matt Carrington
Contributing Designer: Kim Saar Richardson
Photographs Editor: Kathy Creech
Contributing Photographs Editor: Clover Morell
Manager, Cartographic Services: Wayne K. Pichler
Senior Cartographer: John M. Rejba

Marketing:
Associate Director, School and Library Marketing: Jennifer Parello

CONTENTS

POLITICAL WORLD MAP

The world has 196 independent countries and about 50 dependencies. An independent country controls its own affairs. Dependencies are controlled in some way by independent countries. In most cases, an independent country is responsible for the dependency's foreign relations and defense, and some of the dependency's local affairs. However, many dependencies have complete control of their local affairs.

By 2010, the world's population was nearly 7 billion. Almost all of the world's people live in independent countries. Only about 13 million people live in dependencies.

Some regions of the world, including Antarctica and certain desert areas, have no permanent population. The most densely populated regions of the world are in Europe and in southern and eastern Asia. The world's largest country in terms of population is China, which has more than 1.3 billion people. The independent country with the smallest population is Vatican City, with only about 830 people. Vatican City, covering only 1/6 square mile (0.4 square kilometer), is also the smallest in terms of size. The world's largest nation in terms of area is Russia, which covers 6,601,669 square miles (17,098,242 square kilometers).

Every nation depends on other nations in some way. The interdependence of the entire world and its peoples is called *globalism*. Nations trade with one another to earn money and to obtain manufactured goods or the natural resources that they lack. Nations with similar interests and political beliefs may pledge to support one another in case of war. Developed countries provide developing nations with financial aid and technical assistance. Such aid strengthens trade as well as defense ties.

Nations of the World

Name	Map key
Afghanistan	D 13
Albania	C 11
Algeria	D 10
Andorra	C 10‡
Angola	F 10
Antigua and Barbuda	E 6
Argentina	G 6
Armenia	D 12
Australia	G 16
Austria	C 10
Azerbaijan	D 12
Bahamas	D 6
Bahrain	D 12
Bangladesh	D 14
Barbados	E 7
Belarus	C 11
Belgium	C 10
Belize	E 5
Benin	E 10
Bhutan	D 14
Bolivia	F 6
Bosnia-Herzegovina	C 10
Botswana	G 11
Brazil	F 7
Brunei	E 15
Bulgaria	C 11
Burkina Faso	E 9
Burundi	F 11
Cambodia	E 15
Cameroon	E 10
Canada	C 4
Cape Verde	E 8
Central African Republic	E 10
Chad	E 10
Chile	G 6
China	D 14
Colombia	E 6
Comoros	F 12
Congo, Democratic Republic of the	F 11
Congo, Republic of the	F 10
Costa Rica	E 5
Côte d'Ivoire	E 9
Croatia	C 10
Cuba	D 5
Cyprus	D 11
Czech Republic	C 10
Denmark	C 10
Djibouti	E 12
Dominica	E 6
Dominican Republic	E 6
East Timor	F 16
Ecuador	F 6
Egypt	D 11
El Salvador	E 5
Equatorial Guinea	E 10
Eritrea	E 12
Estonia	C 11
Ethiopia	E 11
Federated States of Micronesia	E 17
Fiji	F 1
Finland	B 11
France	C 10
Gabon	F 10
Gambia	E 9
Georgia	C 12
Germany	C 10
Ghana	E 9
Greece	D 11
Grenada	E 6
Guatemala	E 5
Guinea	E 9
Guinea-Bissau	E 9

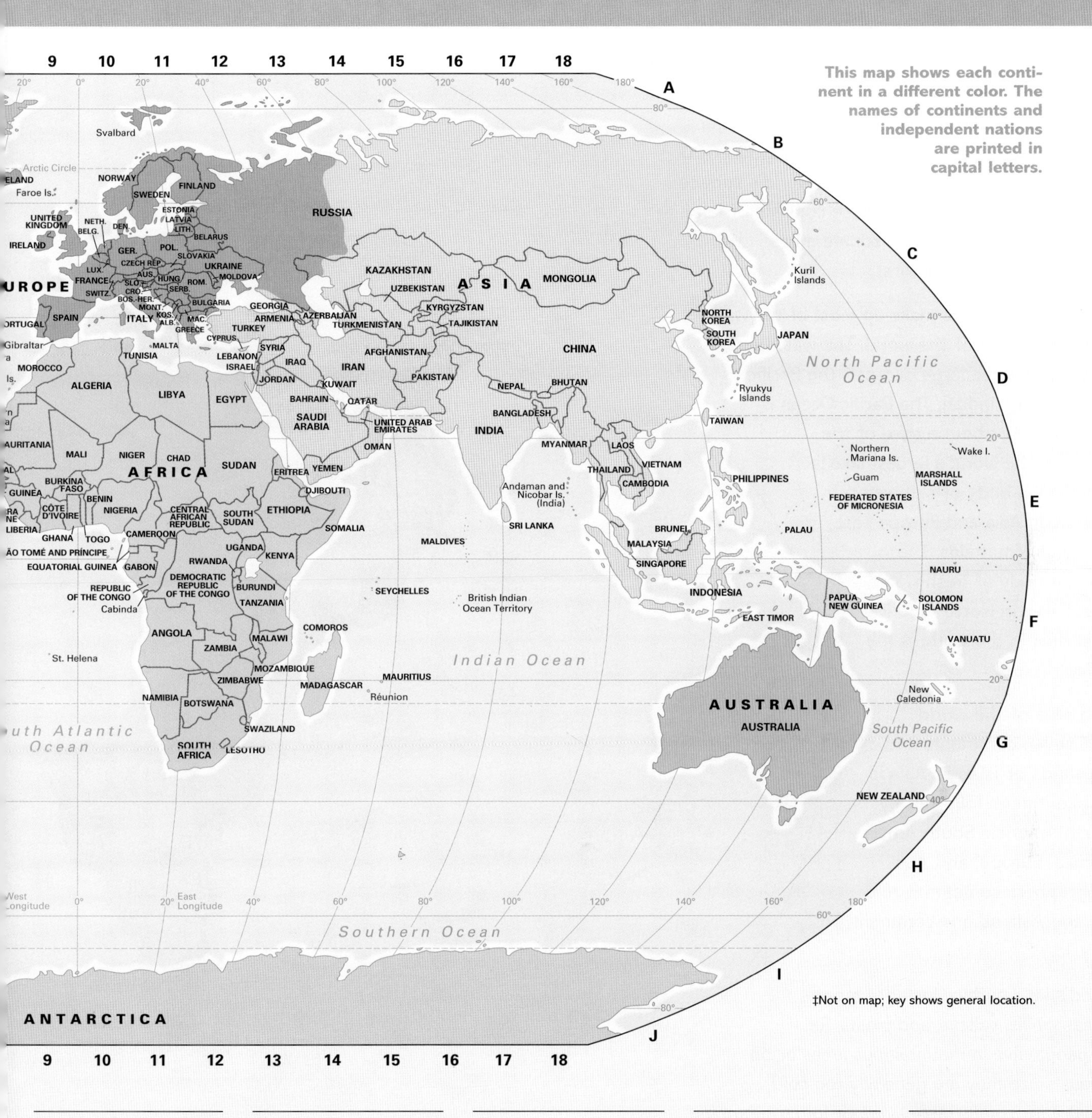

This map shows each continent in a different color. The names of continents and independent nations are printed in capital letters.

‡Not on map; key shows general location.

Name	Map key
Guyana	E 7
Haiti	E 6
Honduras	E 5
Hungary	C 10
Iceland	B 9
India	D 13
Indonesia	F 16
Iran	D 12
Iraq	D 12
Ireland	C 9
Israel	D 11
Italy	C 10
Jamaica	E 6
Japan	D 16
Jordan	D 11
Kazakhstan	C 13
Kenya	E 11
Kiribati	F 1
Korea, North	C 16
Korea, South	D 16
Kosovo	C 11
Kuwait	D 12
Kyrgyzstan	C 13
Laos	E 15
Latvia	C 11
Lebanon	D 11
Lesotho	G 11
Liberia	E 9
Libya	D 10
Liechtenstein	C 10‡
Lithuania	C 11
Luxembourg	C 10
Macedonia	C 11
Madagascar	F 12
Malawi	F 11
Malaysia	E 15
Maldives	E 13
Mali	E 9
Malta	D 10
Marshall Islands	E 18
Mauritania	D 9
Mauritius	G 12
Mexico	D 4
Moldova	C 11
Monaco	C 10‡
Mongolia	C 15
Montenegro	C 10
Morocco	D 9
Mozambique	F 11
Myanmar	D 14
Namibia	G 10
Nauru	F 18
Nepal	D 14
Netherlands	C 10
New Zealand	G 18
Nicaragua	E 5
Niger	E 10
Nigeria	E 10
Norway	B 10
Oman	E 12
Pakistan	D 13
Palau	E 16
Panama	E 5
Papua New Guinea	F 17
Paraguay	G 7
Peru	F 6
Philippines	E 16
Poland	C 10
Portugal	D 9
Qatar	D 12
Romania	C 11
Russia	C 13
Rwanda	F 11
St. Kitts and Nevis	E 6
St. Lucia	E 6
St. Vincent and the Grenadines	E 6
Samoa	F 1
San Marino	C 10‡
São Tomé and Príncipe	E 10
Saudi Arabia	D 12
Senegal	E 9
Serbia	C 10
Seychelles	F 12
Sierra Leone	E 9
Singapore	E 15
Slovakia	C 11
Slovenia	C 11
Solomon Islands	F 18
Somalia	E 12
South Africa	G 11
Spain	C 9
Sri Lanka	E 14
Sudan	E 11
Sudan, South	E 11
Suriname	E 7
Swaziland	G 11
Sweden	B 10
Switzerland	C 10
Syria	D 11
Taiwan	D 16
Tajikistan	D 14
Tanzania	F 11
Thailand	E 15
Togo	E 9
Tonga	F 1
Trinidad and Tobago	E 6
Tunisia	D 10
Turkey	D 11
Turkmenistan	D 13
Tuvalu	F 1
Uganda	E 11
Ukraine	C 11
United Arab Emirates	D 12
United Kingdom	C 9
United States	C 4
Uruguay	G 7
Uzbekistan	D 14
Vanuatu	F 18
Vatican City	C 10‡
Venezuela	E 6
Vietnam	E 15
Yemen	E 12
Zambia	F 11
Zimbabwe	G 11

PHYSICAL WORLD MAP

The surface area of the world totals about 196,900,000 square miles (510,000,000 square kilometers). Water covers about 139,700,000 square miles (362,000,000 square kilometers), or 71 percent of the world's surface. Only 29 percent of the world's surface consists of land, which covers about 57,200,000 square miles (148,000,000 square kilometers).

Oceans, lakes, and rivers make up most of the water that covers the surface of the world. The water surface consists chiefly of three large oceans—the Pacific, the Atlantic, and the Indian. The Pacific Ocean is the largest, covering about a third of the world's surface. The world's largest lake is the Caspian Sea, a body of salt water that lies between Asia and Europe east of the Caucasus Mountains. The world's largest body of fresh water is the Great Lakes in North America. The longest river in the world is the Nile in Africa.

The land area of the world consists of seven continents and many thousands of islands. Asia is the largest continent, followed by Africa, North America, South America, Antarctica, Europe, and Australia. Geographers sometimes refer to Europe and Asia as one continent called Eurasia.

The world's land surface includes mountains, plateaus, hills, valleys, and plains. Relatively few people live in mountainous areas or on high plateaus since they are generally too cold, rugged, or dry for comfortable living or for crop farming. The majority of the world's people live on plains or in hilly regions. Most plains and hilly regions have excellent soil and an abundant water supply. They are good regions for farming, manufacturing, and trade. Many areas unsuitable for farming have other valuable resources. Mountainous regions, for example, have plentiful minerals, and some desert areas, especially in the Middle East, have large deposits of petroleum.

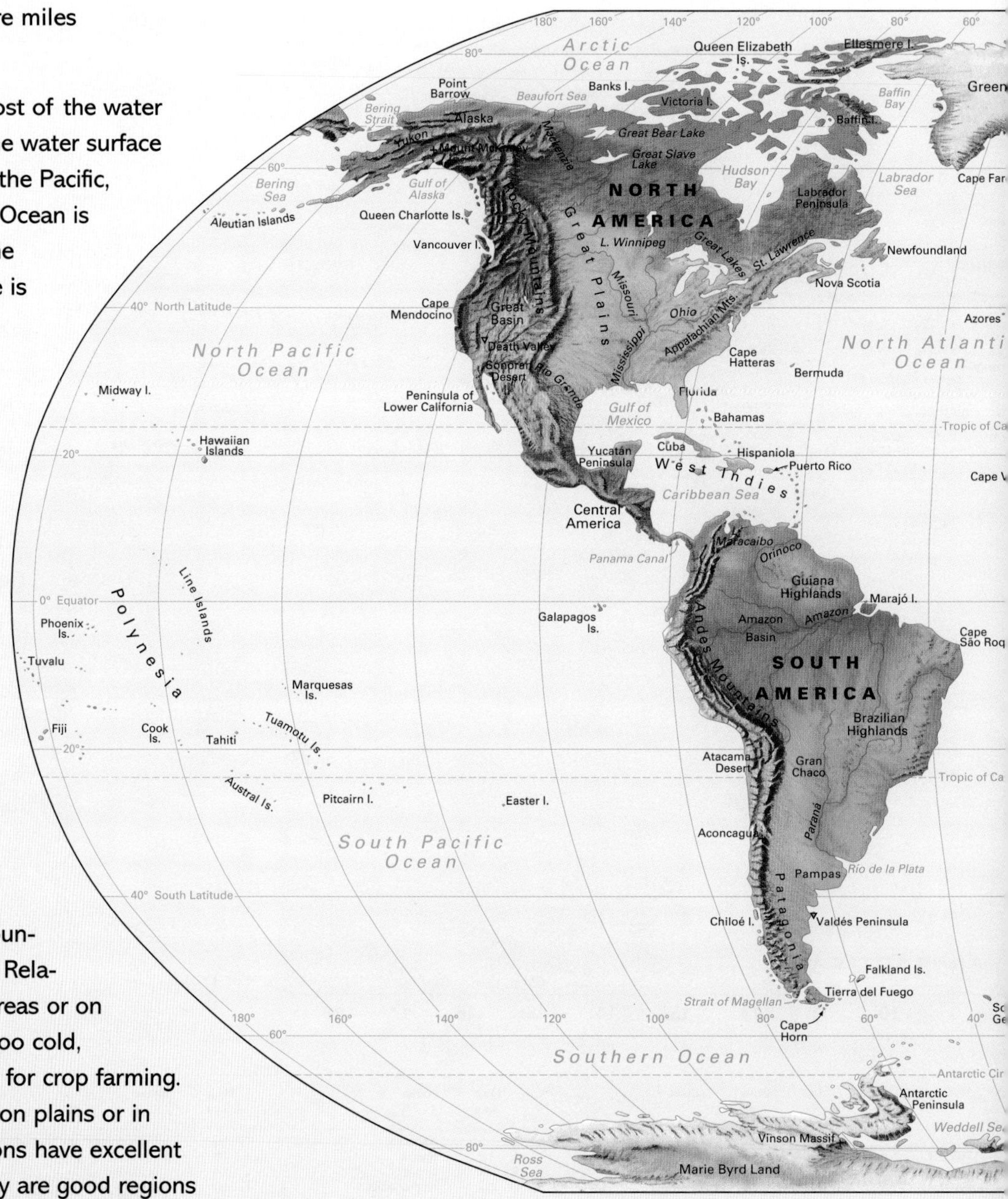

This map shows the world's chief physical features. Areas shown in shades of green generally have fertile soil and sufficient rainfall. Most of the world's people live in these areas.

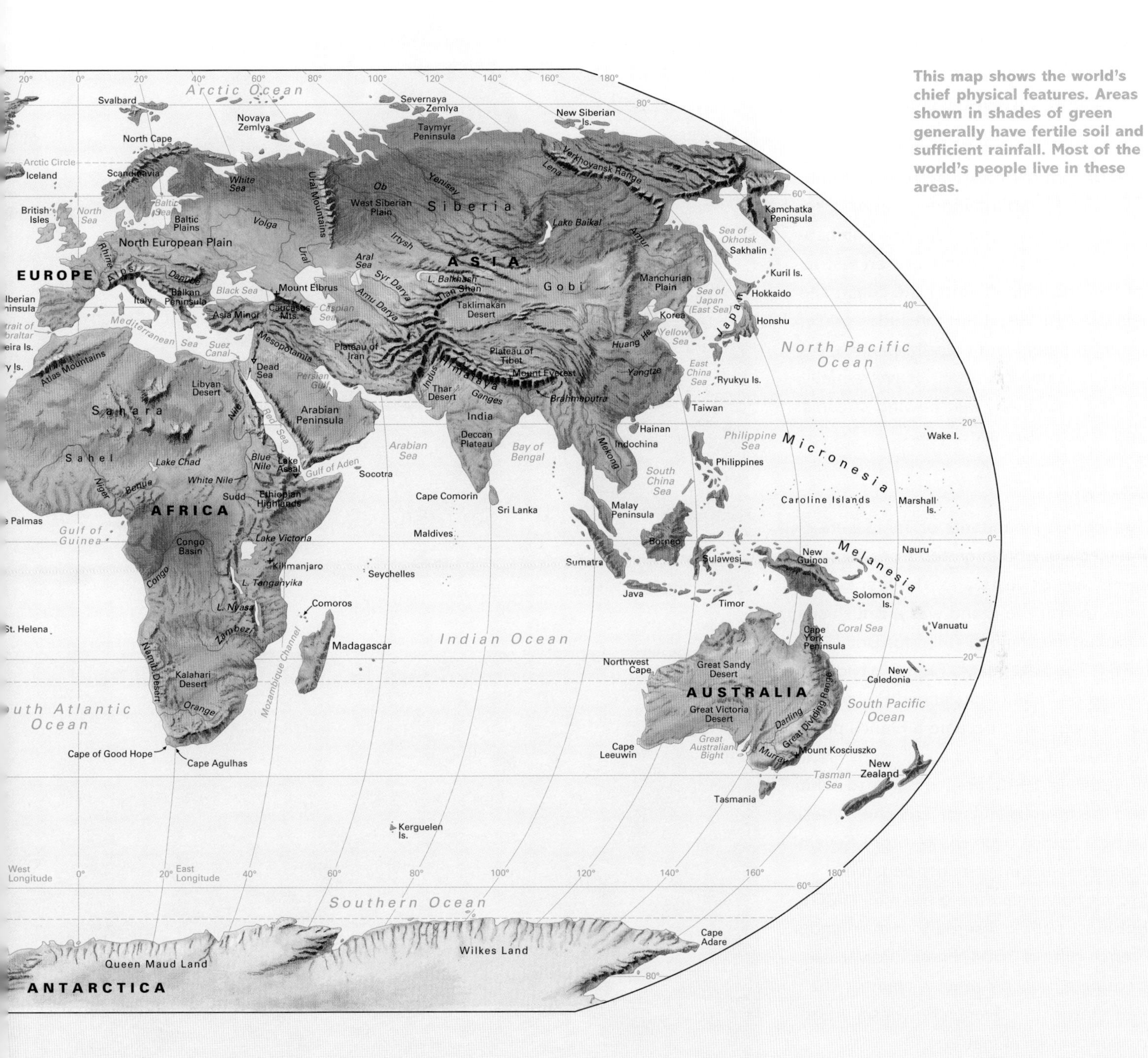

TAHITI

The South Pacific island of Tahiti is one of the 14 Society Islands. It is the largest island in French Polynesia, a French possession made up of several island groups. Tahiti covers 402 square miles (1,041 square kilometers). Temperatures in this tropical climate hover near 81° F (27° C) throughout the year.

A broken coral reef surrounds Tahiti. The island's mountainous interior is so steep that it is almost entirely uninhabited. Most of the people live on a strip of flat, fertile land along the coast. The island's lush vegetation includes graceful coconut palms and colorful banana, orange, and papaya trees. Heavy rainfall helps create many fast-flowing streams and spectacular waterfalls.

History

The earliest inhabitants of Tahiti were Polynesians who migrated from Asia thousands of years ago. The first European to visit the island was a British sea captain who claimed Tahiti for Britain in 1767. The next year, a French navigator landed on Tahiti and claimed the island for France. Tahiti became a French protectorate in 1842 and a colony of France in 1880. In 1946, France declared Tahiti and the other islands of French Polynesia to be a French overseas territory. Several independence movements began in French Polynesia during the mid-1900's, but most people wanted to remain under French rule. In 2004, French Polynesia became a French overseas country. The new status gave French Polynesia more power over local matters.

Papeete, on Tahiti's north coast, is the capital and largest community of French Polynesia, as well as the chief port of Tahiti. Many Tahitians live in or near Papeete and work in the tourist industry.

Raftlike boats called *catamarans* once carried people on long voyages between Polynesian islands. Most of the islands of Polynesia were settled later than those of Melanesia and Micronesia.

A young Tahitian girl, hibiscus flowers in her hair, has the gentle look that Paul Gauguin idealized in his paintings of South Sea Islands people.

Tahiti, the largest of the 14 Society Islands, occupies a total land area of 402 square miles (1,041 square kilometers). A coral reef surrounds the island, and steep mountains rise in the island's interior.

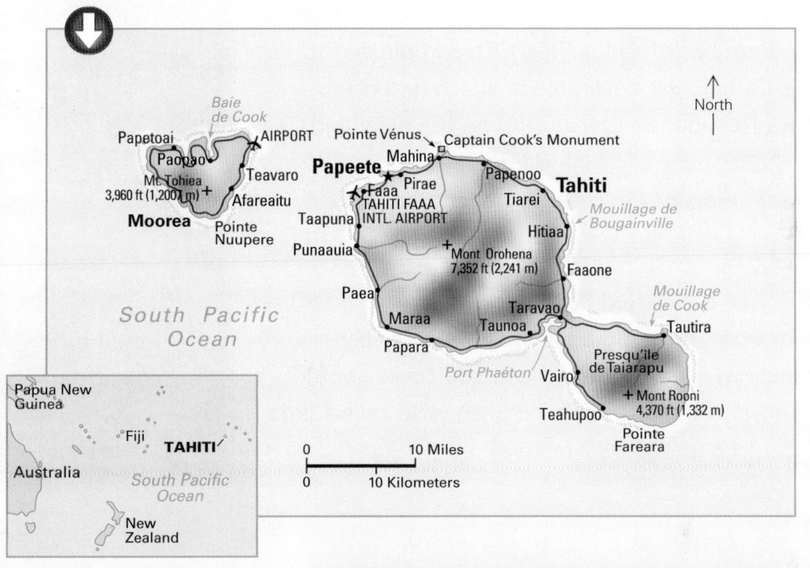

Papeete, Tahiti's largest city and chief port, is also the capital of French Polynesia. The voters of French Polynesia elect a territorial assembly, which in turn elects the president of French Polynesia. The islands also send representatives to the French Parliament.

Tahiti gained worldwide fame as a tropical paradise through the works of many artists and writers. The paintings of the French artist Paul Gauguin captured Tahiti's gentle people, lush beauty, and peaceful atmosphere. Such authors as Herman Melville and James Michener of the United States, as well as Robert Louis Stevenson of Scotland, wrote glowing descriptions of the island. These works helped make Tahiti popular with tourists.

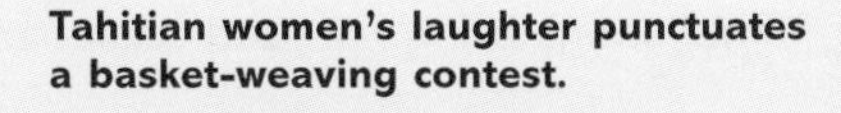

Tahitian women's laughter punctuates a basket-weaving contest.

TAIWAN

In beautiful Taroko Gorge, a Buddhist shrine stands above a small waterfall. The swift waters of the small Li-wu River carved the gorge through Taiwan's mountains.

The mountainous island nation of Taiwan lies in the South China Sea, about 90 miles (140 kilometers) off the coast of mainland China. The name *Taiwan* means *terraced bay* in Chinese.

Taiwan's political status has been a subject of dispute since the mid-1900's, when Chinese Communists and Chinese Nationalists clashed over control of China. After the Communists seized power on the Chinese mainland in 1949, the Nationalists, under the leadership of Chiang Kai-shek, moved to Taiwan. The Nationalists declared Taipei, at the northern end of Taiwan, the capital of the Republic of China (ROC). Chiang refused to recognize China's Communist government, which had established the People's Republic of China (PRC) on the mainland. Today, the People's Republic of China claims Taiwan is a PRC province, but Taiwan remains effectively independent.

Land

Taiwan's wild, forested beauty led Portuguese sailors in 1590 to name it *Ilha Formosa*, meaning *beautiful island*. Thickly forested mountains run from north to south and cover about half of Taiwan. Taiwan's highest peak, Yü Shan (also called Jade Mountain) rises 13,113 feet (3,997 meters).

Electronics are among Taiwan's leading export products. Taiwan is one of the world's largest producers of computer equipment.

At many places along the island's eastern coast, the mountains drop sharply to the sea. Short, swift rivers have cut gorges through the mountains, creating a scenic landscape of sheer cliffs and wide, sandy beaches. In western Taiwan, the mountains descend gently into rolling hills and a wide plain. Most of Taiwan's people live west of the mountains, and all of the island's large cities are located there.

The government of Taiwan also rules the Pescadores islands, located just off the island's west coast, and the Quemoy and Matsu groups, which lie very close to mainland China. The landscape of the Pescadores is much like that of Taiwan, but the Quemoy and Matsu island groups are not as mountainous.

Taiwan has a subtropical climate, with hot, humid summers and mild winters. Abundant rain falls in Taiwan, averaging more than 100 inches (250 centimeters) annually. Summer monsoons bring strong winds and rain, while winter monsoons bring rain and cooler weather to the north.

Taipei 101, in Taipei, is one of the tallest buildings in the world. The 101-story structure towers above the business district and dominates the city's skyline.

Today, Taiwan has one of the strongest economies in Asia. It depends heavily on manufacturing and foreign trade, and it exports more than it imports. Taiwan's leading trade partners include China, Japan, and the United States. Many Taiwanese live in China or regularly travel to China for work.

Taiwan's factories produce cement, clothing and textiles, computer equipment, electronics, iron and steel, machinery, motor vehicles, plastic goods, processed foods, televisions, and toys. Most of these goods are exported. Taiwan also has some mining of limestone, marble, natural gas, petroleum, and sulfur.

Most of the country is too mountainous for farming, and only about 25 percent of the land is suitable for agriculture. Taiwan's farmers have terraced many hills to provide more fields for growing rice, which is the main crop. By using fertilizers, they can harvest two or three crops a year from the same land. Other crops include bamboo, bananas, betel nuts, citrus fruits, corn, peanuts, pears, pineapples, sugar cane, sweet potatoes, and tea. Taiwan also produces and exports orchids and other flowers. Farmers raise cattle, chickens, ducks, and hogs.

The fishing industry catches squid and tuna. Eels and other freshwater fish are raised and caught in inland ponds.

A farmer tends a field of grain on Taiwan's eastern coast. Although agriculture has declined in importance relative to industrial and service activities, it is still a vital part of Taiwan's economy.

Temperatures average about 80° F (27° C) in the summer and 65° F (18° C) in winter.

Typhoons plague Taiwan almost every year. The storms, which are among the most severe in the world, cause flooding and major damage to buildings and crops.

Taiwan's economic miracle

In 1949, when Taiwan's Nationalist government was established, the country's leaders concentrated on developing the island's economy. Agricultural growth was achieved by land reforms, innovative farming techniques, and aid and advice from the United States. Industrial productivity benefited from the initial low cost of labor and the hard-working people of Taiwan. Later, labor-saving devices and efficient organization helped increase productivity.

TAIWAN TODAY

Beginning in the early 1900's, two groups—the Nationalist Party and the Communists—vied for control of China. In 1946, their struggle escalated into civil war, which ended in 1949 with a Communist victory. The Nationalist leaders, under Chiang Kai-shek, withdrew to Taiwan and set up a rival government. They named the island the Republic of China (ROC) and made Taipei the capital.

Today, both the Communist Chinese government and the Nationalist government of Taiwan claim to be the legitimate ruler of all China, and both governments consider Taiwan a province of China. In the 1950's and 1960's, Taiwan represented China in the United Nations (UN), and the United States supported the country both economically and militarily. However, in 1971, the UN expelled the Nationalists and admitted Communist China. The United States ended its official diplomatic relations with Taiwan in 1978. However, the two countries agreed to carry on unofficial relations through nongovernmental agencies, and trade between the two continues to thrive.

Government

The ROC government of Taiwan is based on a constitution adopted in 1946 on the Chinese mainland. Since the 1980's, the Constitution has been amended several times, and Taiwan's government has become democratic.

Taiwan has five branches of government—executive, legislative, judicial, control, and examination. Each branch, or *yuan,* is headed by a president. Taiwan's head of state is the president of the Republic of China, who is elected by the people to a four-year term. The president appoints a premier to head the Executive Yuan, which carries out the operations of the government. The premier is also known as the president of the Executive Yuan. The elected Legislative Yuan makes most of Taiwan's laws. It also approves the appointment of the premier.

Politics

Chiang Kai-shek was president of Taiwan until his death in 1975. Chiang's son Chiang Ching-kuo, who had

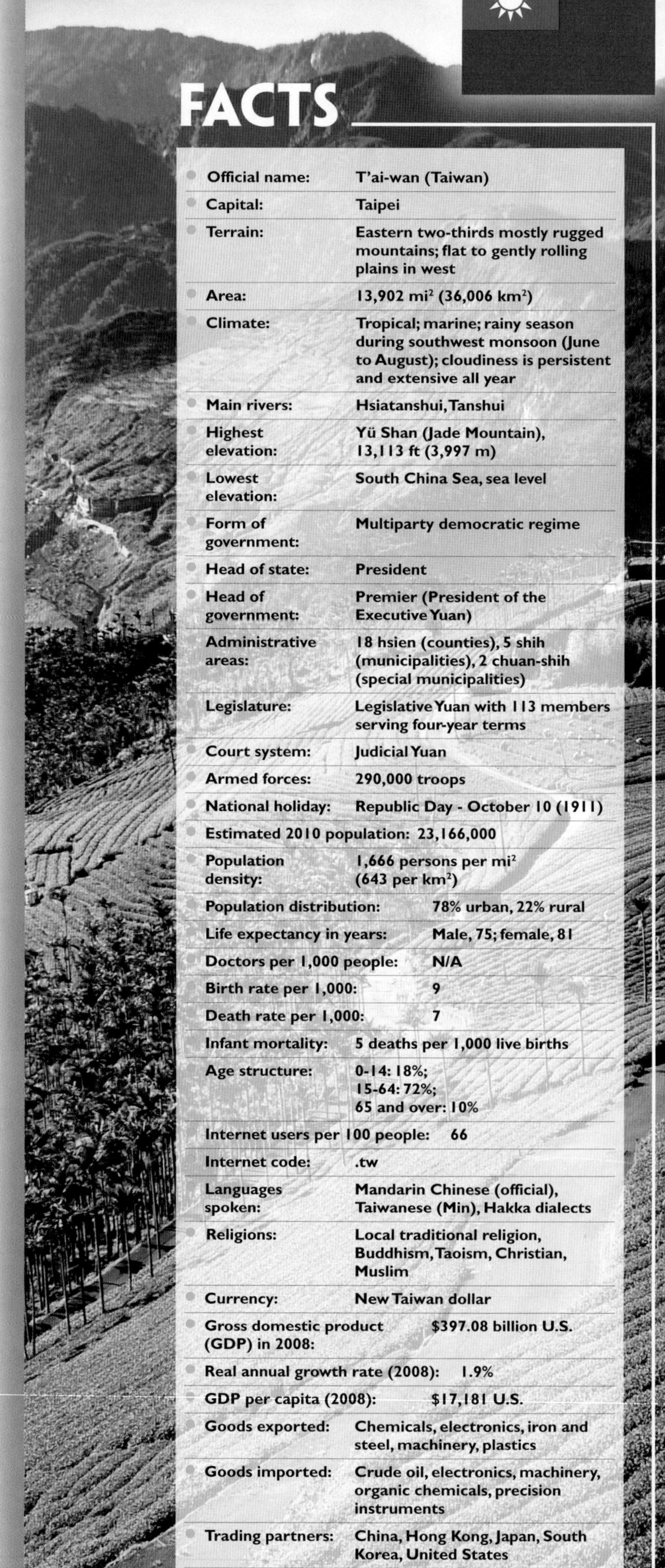

FACTS

Official name:	T'ai-wan (Taiwan)
Capital:	Taipei
Terrain:	Eastern two-thirds mostly rugged mountains; flat to gently rolling plains in west
Area:	13,902 mi² (36,006 km²)
Climate:	Tropical; marine; rainy season during southwest monsoon (June to August); cloudiness is persistent and extensive all year
Main rivers:	Hsiatanshui, Tanshui
Highest elevation:	Yü Shan (Jade Mountain), 13,113 ft (3,997 m)
Lowest elevation:	South China Sea, sea level
Form of government:	Multiparty democratic regime
Head of state:	President
Head of government:	Premier (President of the Executive Yuan)
Administrative areas:	18 hsien (counties), 5 shih (municipalities), 2 chuan-shih (special municipalities)
Legislature:	Legislative Yuan with 113 members serving four-year terms
Court system:	Judicial Yuan
Armed forces:	290,000 troops
National holiday:	Republic Day - October 10 (1911)
Estimated 2010 population:	23,166,000
Population density:	1,666 persons per mi² (643 per km²)
Population distribution:	78% urban, 22% rural
Life expectancy in years:	Male, 75; female, 81
Doctors per 1,000 people:	N/A
Birth rate per 1,000:	9
Death rate per 1,000:	7
Infant mortality:	5 deaths per 1,000 live births
Age structure:	0-14: 18%; 15-64: 72%; 65 and over: 10%
Internet users per 100 people:	66
Internet code:	.tw
Languages spoken:	Mandarin Chinese (official), Taiwanese (Min), Hakka dialects
Religions:	Local traditional religion, Buddhism, Taoism, Christian, Muslim
Currency:	New Taiwan dollar
Gross domestic product (GDP) in 2008:	$397.08 billion U.S.
Real annual growth rate (2008):	1.9%
GDP per capita (2008):	$17,181 U.S.
Goods exported:	Chemicals, electronics, iron and steel, machinery, plastics
Goods imported:	Crude oil, electronics, machinery, organic chemicals, precision instruments
Trading partners:	China, Hong Kong, Japan, South Korea, United States

Chiang Kai-shek Memorial Hall in Taipei honors the Nationalist Party leader who ruled Taiwan from 1949 to 1975. Chiang Kai-shek's son, Chiang Ching-kuo, who became president of Taiwan in 1978, laid the groundwork for a constitutional democracy and worked to reduce the Nationalist Party's power.

served as Taiwan's prime minister since 1972, was elected president in 1978. He took steps to democratize Taiwan by ending Taiwan's martial law, which the government had imposed in the 1950's. He also reformed Taiwan's political system by legalizing opposition political parties.

When Chiang Ching-kuo died in 1988, Vice President Lee Teng-hui succeeded him as president and pledged to continue the reforms. Lee was the first president born in Taiwan rather than the Chinese mainland, and he was popular with the island's native Taiwanese. The native Taiwanese are those people whose families had lived on the island before the massive migration of people fleeing the Communist take-over of mainland China in 1949. To many of Taiwan's people, Lee symbolized democracy as well as a shift in power from Chinese mainlanders to the Taiwanese.

Lee was reelected in 1996, in the first direct, democratic election for a national leader in Taiwan's history. In 2000, Chen Shui-bian of the Democratic Progressive Party was elected president, ending 50 years of Nationalist Party rule. In 2008, Nationalist Party candidate and former Taipei mayor Ma Ying-jeou succeeded Chen as president. He was reelected in 2012.

Taiwan is separated from mainland China by the narrow Taiwan Strait. In addition to Taiwan, the Nationalist government also governs several islands in the Taiwan Strait, including the Matsu, Pescadores, and Quemoy island groups.

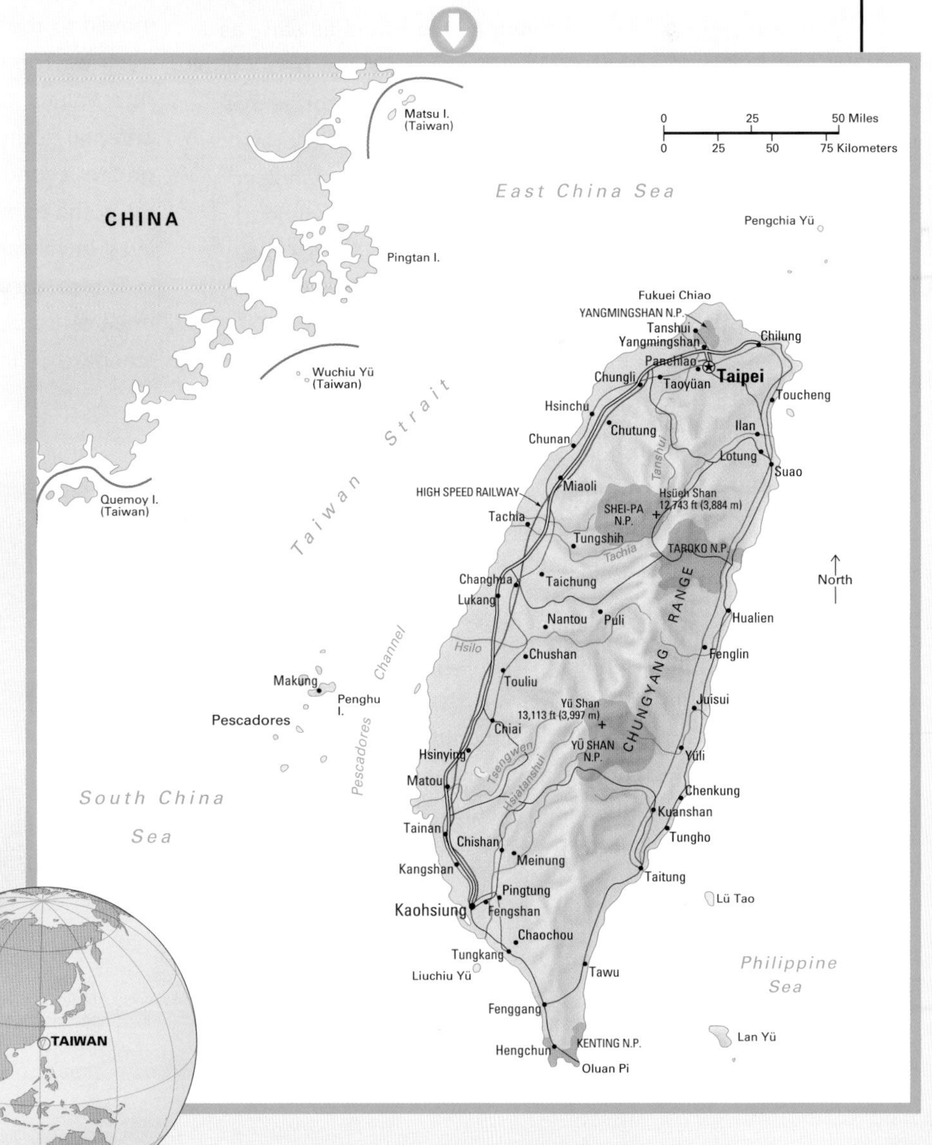

PEOPLE AND CULTURE

The Ami, an aboriginal group in Taiwan, celebrate a harvest festival with a traditional dance. The Ami people, along with two other groups, account for most of Taiwan's native population.

Almost all of Taiwan's more than 23 million people live on the coastal plain that makes up the western third of the island. With more than 1,600 people per square mile (600 per square kilometer), Taiwan is one of the most densely populated countries in the world.

Origins and history

A small percentage of Taiwan's people are non-Chinese native people, sometimes called aborigines. Their ancestors lived on the island before Chinese settlers arrived. Taiwan's aborigines are related to the original inhabitants of other Pacific islands. They live mostly in the mountains.

Chinese settlers began arriving on the island as early as the A.D. 500's. But large-scale settlement did not begin until the 1600's, when the Chinese Manchu dynasty conquered Taiwan and made it part of the southern mainland province of Fujian. Taiwan became a separate Chinese province in 1885. About 80 percent of all Taiwanese are descended from Chinese settlers who came to the island before the mid-1900's, mainly from the provinces of Guangdong (Kwangtung) and Fujian.

Japan gained control of Taiwan at the end of the Sino-Japanese War of 1894-1895 and ruled it until 1945, when China regained control after World War II. After the Communist take-over of mainland China in 1949, more than 2 million Chinese people fled to Taiwan.

The people of Taiwan speak various Chinese dialects. Most speak Northern Chinese, also known as Mandarin. In Taiwan, that dialect is called *Guoyu,* meaning *national language.* Another widely spoken dialect is Taiwanese, a form of Minnon, the dialect spoken in the Fujian province of mainland China.

Way of life

In the early 1900's, most of Taiwan's people lived in rural areas. As a result of the agricultural reforms of the early 1950's, however, farming became more efficient, and fewer people were needed to work on the farms. Many rural people then moved to the cities to seek jobs in Taiwan's developing industries. The resulting urban growth weakened traditional rural values, which were based on strong family ties and parental authority. However, urbanization also provided greater economic opportunities and promoted social equality. In the early 2000's, about 80 percent of the people lived in Taiwan's cities and towns.

The country's economic growth brought prosperity to most of its people. Taiwanese farmers also had a higher standard of living than farmers in other Asian countries.

Carved columns and colorful lanterns decorate a temple in Taipei. About half of Taiwan's people practice a traditional religion that involves the worship of gods and goddesses. Buddhism and Taoism are the most common major religions.

Preserving Chinese heritage

After the Nationalists moved to Taiwan in 1949, they took steps to preserve many aspects of traditional Chinese culture. The government launched a program promoting Chinese calligraphy, classical painting, Beijing opera, and folk arts. The government also opened the National Palace Museum in Taipei, which houses an outstanding collection of Chinese art and antiques brought to Taiwan by the Nationalists.

As Taiwan gradually acquired an identity of its own, culture became broader in scope. Sculpture, for example, joined traditional Chinese calligraphy and painting as a popular art form in Taiwan. Nationalism became a common theme among Taiwanese writers. As the government loosened some of its restrictions on artistic expression, many artists felt free to deal with modern, controversial topics.

A food stall serves visitors in Snake Alley, a market open at night in Taipei. The city has several open-air markets that are popular tourist attractions.

CHIANG KAI-SHEK

Chiang Kai-shek stressed Taiwan's Chinese heritage and its links with mainland China. His interests influenced government policy regarding cultural involvement.

In 1949, Chiang Kai-shek led the Nationalist government—and nearly 2 million mainland Chinese—to Taiwan, where he hoped to plan a counterattack against the Communist Chinese. Although they did not like the Chinese mainlanders or the Nationalist government, most Taiwanese accepted Nationalist rule as preferable to being governed by Communist China.

In time, as the hopelessness of reconquering mainland China became evident, the Nationalists began to concentrate their leadership on Taiwan. In the 1950's, Chiang Kai-shek instituted various reforms to help stimulate Taiwan's economy, and, as their island prospered, many Taiwanese put aside their dislike of the Nationalists. In 1978, three years after Chiang Kai-shek's death, his son Chiang Ching-kuo was elected president. Chiang Ching-kuo's democratic reforms and continued economic development of the island made him popular with both the Taiwanese and the mainlanders on Taiwan. Chiang Ching-kuo died in 1988. He was succeeded by Vice President Lee Teng-hui, who continued the democratization process.

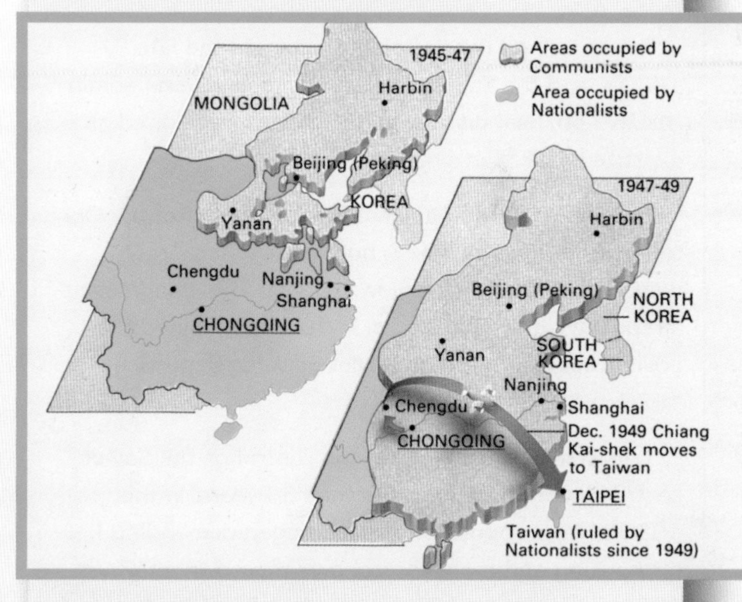

China's civil war was fought from 1946 to 1949 between the Nationalists led by Chiang Kai-shek and the Chinese Communists led by Mao Zedong. During World War II (1939-1945), the Communists established bases in northern China, while the Nationalists held the south. When full-scale civil war erupted in 1946, the superior military tactics of the Communists turned the tide against the Nationalists, who fled to Taiwan in 1949. The United States helped protect Taiwan from Communist Chinese invasion in the early 1950's.

TAJIKISTAN

Tajikistan is a mountainous country bordered by China on the east, Afghanistan in the south, Kyrgyzstan on the north, and Uzbekistan on the west and northwest. The Pamirs, one of the highest mountain ranges in the world, rise to towering heights in the east.

History

The ancestors of present-day Tajiks were Persians from the Achaemenid Empire, which was centered in what it now Iran. They arrived as early as the 500's B.C. and created thriving communities of farmers and merchants. Many conquering armies passed through the region, often following the Amu Darya or the Syr Darya—two of the most important rivers in central Asia. These conquerors included Macedonians led by Alexander the Great, Seleucids, Bactrians, Chinese nomads, Kushans, Sassanians, and White Huns. Arab armies introduced Islam to the region in the middle of the A.D. 600's. Turks conquered the area in the 700's, Mongols seized it in the 1200's, and Uzbeks ruled it from the 1500's to 1800's.

Russia gained control of the region in the late 1800's, and the Tajiks were split among several administrative-political districts. The Russians developed and exploited Tajikistan's natural resources.

The Tajiks rebelled against Russian rule after the October Revolution of 1917, but Russia's Red Army gained control of the region in 1921. The area became a republic within Uzbekistan in 1924 and joined the Soviet Union as the Tajik Soviet Socialist Republic in 1929.

Tajikistan remained under the control of the Soviet central government for more than 60 years. In the 1980's, Tajik opposition groups began demanding better housing and more control over their own affairs. In 1989, Tajik replaced Russian as the official language of the republic. In 1990, Tajikistan declared that its laws overruled Soviet laws. Finally, in September 1991, Tajikistan declared its independence. The Soviet Union was formally dissolved later that year.

After Tajikistan achieved independence, violence broke out between the Communist government and an

FACTS

Official name:	Jumhurii Tojikiston (Republic of Tajikistan)
Capital:	Dushanbe
Terrain:	Pamir and Alay mountains dominate landscape; valleys in north and southwest
Area:	55,251 mi² (143,100 km²)
Climate:	Midlatitude continental, hot summers, mild winters; semiarid to polar in Pamir Mountains
Main rivers:	Amu Darya, Syr Darya, Zeravshan
Highest elevation:	Pik Imeni Ismail Samani, 24,590 ft (7,495 m)
Lowest elevation:	Syr Darya river at northwestern border, 980 ft (300 m)
Form of government:	Republic
Head of state:	President
Head of government:	Prime minister
Administrative areas:	2 viloyatho (provinces), 1 viloyati mukhtori (autonomous province)
Legislature:	Majlisi Oli (Supreme Assembly) consisting of the Majlisi Milliy (National Assembly) with 34 members, and the Majlisi Namoy-andagon (Assembly of Representatives) with 63 members, all serving five-year terms
Court system:	Supreme Court
Armed forces:	8,800 troops
National holiday:	Independence Day - September 9 (1991)
Estimated 2010 population:	7,389,000
Population density:	134 persons per mi² (52 per km²)
Population distribution:	76% rural, 24% urban
Life expectancy in years:	Male, 63; female, 69
Doctors per 1,000 people:	2.0
Birth rate per 1,000:	27
Death rate per 1,000:	6
Infant mortality:	57 deaths per 1,000 live births
Age structure:	0-14: 38%; 15-64: 58%; 65 and over: 4%
Internet users per 100 people:	7
Internet code:	.tj
Languages spoken:	Tajik (official), Russian
Religions:	Sunni Muslim 85%, Shiah Muslim 5%, other 10%
Currency:	Somoni
Gross domestic product in 2008:	$5.13 billion U.S. (GDP)
Real annual growth rate (2008):	7.9%
GDP per capita (2008):	$704 U.S.
Goods exported:	Mostly aluminum and aluminum products. Also: cotton, fruits
Goods imported:	Aluminum oxide, electric power, food, natural gas, petroleum
Trading partners:	China, Russia, Netherlands, Turkey, Uzbekistan

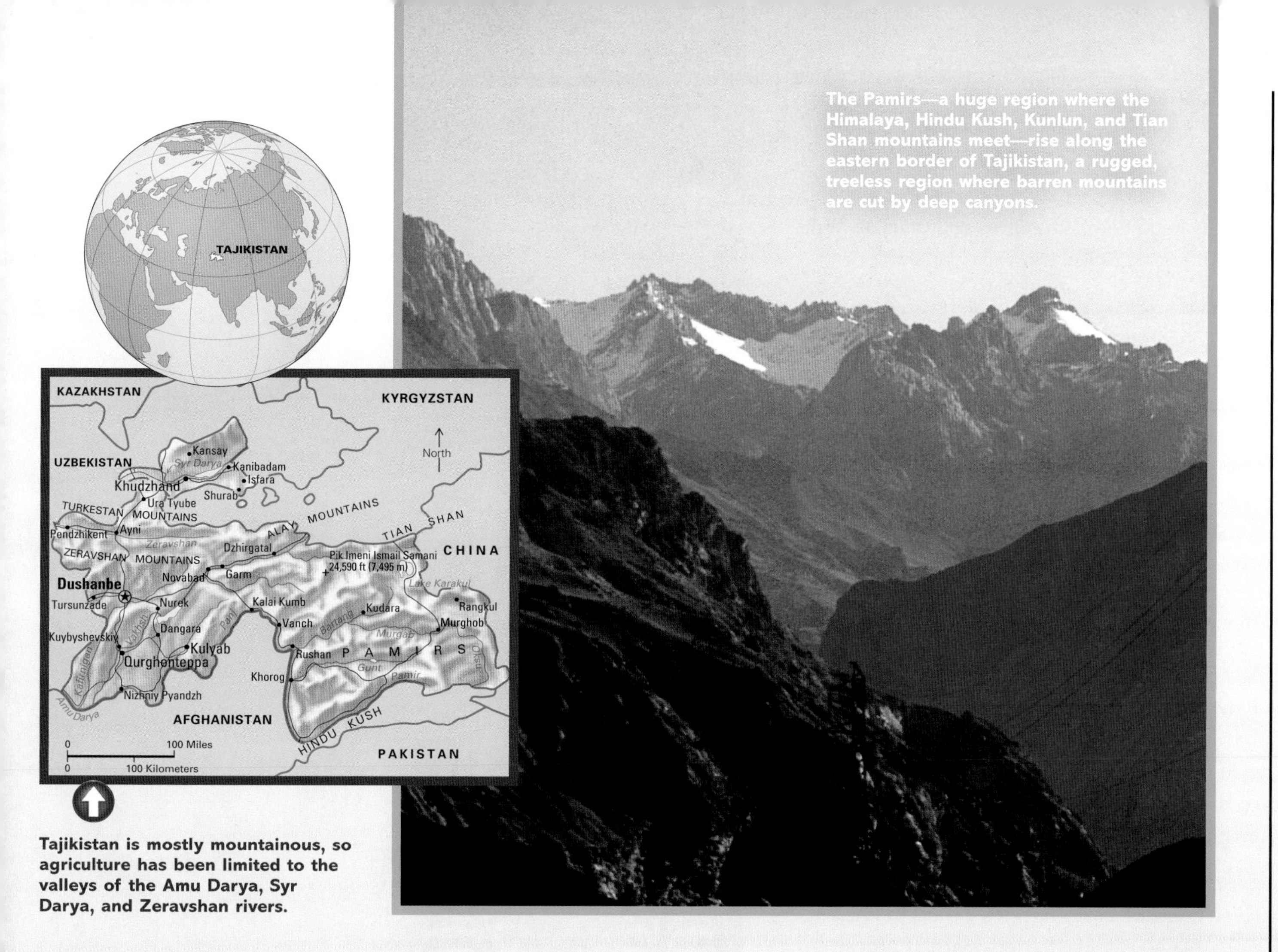

Tajikistan is mostly mountainous, so agriculture has been limited to the valleys of the Amu Darya, Syr Darya, and Zeravshan rivers.

The Pamirs—a huge region where the Himalaya, Hindu Kush, Kunlun, and Tian Shan mountains meet—rise along the eastern border of Tajikistan, a rugged, treeless region where barren mountains are cut by deep canyons.

alliance of prodemocracy and Islamic forces. Tens of thousands of people died, and many more lost their homes in the fighting. The country adopted a new constitution in 1994. A peace agreement ended the civil war in 1997.

Economy and people

Tajikistan's economy is based mostly on agriculture, livestock production, mining, and the processing of raw materials. Cotton is the main agricultural product. Farmers also grow onions, potatoes, tomatoes, wheat, and various fruits. They raise cattle, chickens, goats, horses, and Karakul sheep. The main agricultural areas lie in the country's southwest and north. The huge Nurek Dam on the Vakhsh River provides water for irrigation and hydroelectric power for industry.

Tajikistan's manufactured products include aluminum, chemicals, food products, and textiles. The country's mines produce antimony, coal, gold, lead, mercury, natural gas, petroleum, salt, silver, and zinc. Unemployment is high, however. Many Tajiks take temporary or permanent jobs in other countries, especially Russia. Money sent back to Tajikistan from these migrant workers has become a major economic resource.

Most of the people in Tajikistan are ethnic Tajiks. Their language is much like Farsi, the chief language of Iran. Uzbeks are the largest minority. Other groups include Kyrgyz, Russians, and Turkmen. Most Tajiks belong to the Sunni branch of Islam. Small communities of other Muslims groups—Shiites and Ismaili Khoja Muslims—live in the mountains.

About three-fourths of the country's people live in rural areas. Families tend to be large, and several generations may live together. People wear both Western-style and traditional clothing. Traditional garments include cotton trousers and robes for men and embroidered silk dresses for women. Both men and women wear embroidered skullcaps.

The country of Tanzania in eastern Africa is a land of fascinating wildlife and spectacular scenery. An incredible variety of animals roam across the country's national parks. This wildlife includes huge herds of zebras and gnus, as well as elephants, giraffes, leopards, and lions.

The vast Serengeti Plain of northwestern Tanzania is home to many of these animals. Africa's highest mountain, the majestic, snow-capped Kilimanjaro, rises 19,340 feet (5,895 meters) in northern Tanzania, while Lake Tanganyika, the longest freshwater lake in the world, stretches for 420 miles (680 kilometers) along the country's western border. The northern mountains and Lake Tanganyika are part of the Great Rift Valley, which runs through eastern Africa.

In the Olduvai Gorge area of northern Tanzania, scientists have found the remains of some of the earliest known human settlements. The prehistoric beings who lived in what is now Tanzania were hunters and gatherers. About 3,000 years ago, people began to farm and raise livestock in the area.

Arab traders began to settle along the Indian Ocean coast of eastern Africa during the A.D. 700's. Many of these traders married African women. The Arab-African families and their settlements produced the Swahili culture. Major trading centers developed on Zanzibar, an island off the coast, and on other islands.

Portuguese traders gained control of the coast in the early 1500's, but local rebellions pushed them out late in the 1600's. Arabs then took control of Zanzibar and developed an extensive trade with the Nyamwezi and Yao peoples, who lived on the mainland. The caravans of these two groups brought gold, ivory, and slaves from the interior to the coast in exchange for ceramics, cloth, and glassware from Asia. The Arabs bought and sold thousands of African slaves, and Zanzibar became a major slave market.

During the 1800's, Europeans again tried to win control of Tanzania, and explorers and missionaries traveled deep into the interior. In the 1880's, Germany took control of the mainland and forced many Africans to work on planta-

TANZANIA

tions. The Africans revolted in 1905 in an incident called the Maji Maji rebellion, but the rebellion failed and thousands were killed.

Meanwhile, the United Kingdom had made Zanzibar and another island, Pemba, a protectorate. The British gradually took over the powers of the Arab sultan who had ruled Zanzibar. When Germany was defeated in World War I (1914-1918), the United Kingdom gained control of the mainland and named it Tanganyika. Thousands of Indians began to immigrate to the region.

In order to prepare Tanganyika for independence after World War II (1939-1945), the British wanted to set up a political system in which Europeans, Asians, and Africans would have equal representation. But the Africans argued that such a system would deny them their rights because they made up the vast majority of the population. In 1954, the Africans formed the Tanganyika African National Union (TANU).

Led by Julius Nyerere and others, TANU won independence with majority rule for Tanganyika in 1961. The next year, Nyerere was elected president of Tanganyika. In 1963, Zanzibar was granted its independence. In April 1964, the two regions joined to form the United Republic of Tanzania, and Nyerere was again elected president.

Nyerere chose the course of *socialism* for Tanzania. He based the country's economic system on *ujamaa,* a traditional African concept of cooperation and self-reliance. But a war with Uganda in the late 1970's exhausted Tanzania's resources at a time when the country's economy was already suffering from high fuel prices.

In the 1980's, trade deficits and debt added to the economic problems, and the government began to turn away from strict socialism. In 1985, Nyerere resigned. His successor, Ali Hassan Mwinyi, continued to decrease the government's control of business in the 1990's.

During the 1990's and early 2000's, Tanzania's trade expanded rapidly while its debt declined. More people began using automobiles, the Internet, and cellular telephones. Some people became wealthy, but poverty became more widespread.

TANZANIA TODAY

The United Republic of Tanzania is a poor country with a relatively large land area. Tanzania's population consists mainly of black Africans. The rest are people of Asian or European descent. Dar es Salaam is Tanzania's largest city, and Dodoma is the capital.

Transportation and communication services in Tanzania are largely underdeveloped. Few roads are paved, and most are poorly maintained. However, Tanzania operates rail lines that connect Dar es Salaam with Zambia to the west, and three international airports serve the country. Tanzania has a number of newspapers, and Dar es Salaam broadcasts a number of radio and television stations throughout the country.

The nation's economic problems have caused shortages of school supplies, and only about half of the children go to elementary school. However, most adults in Tanzania can read and write.

Government

A president heads Tanzania's government. The people choose the president in elections every five years. A cabinet, which includes the vice president and a prime minister, assists the president. The National Assembly is Tanzania's lawmaking body. Most members of parliament are elected by the people every five years. Some members are appointed. The Zanzibar administration has its own president and House of Representatives.

Until 1992, Tanzania was a single-party state. The Chama Cha Mapinduzi (CCM) or Revolutionary Party was the only legal party. In 1995, Tanzania held its first multiparty general elections. However, the former single party of the CCM was elected. The CCM remained in control after elections in 2000, 2005, and 2010.

Land

Tanzania borders the Indian Ocean for about 500 miles (800 kilometers). Like other tropical shores, Tanzania's coastal lowland strip is lined with mangrove swamps and coconut palm groves. In addition to Zan-

FACTS

Official name:	United Republic of Tanzania
Capital:	Dodoma
Terrain:	Plains along coast; central plateau; highlands in north, south
Area:	364,900 mi² (945,087 km²)
Climate:	Varies from tropical along coast to temperate in highlands
Main rivers:	Rufiji, Pangani, Ruvuma, Wami
Highest elevation:	Kilimanjaro, 19,340 ft (5,895 m)
Lowest elevation:	Indian Ocean, sea level
Form of government:	Republic
Head of state:	President
Head of government:	President
Administrative areas:	26 regions
Legislature:	Bunge (National Assembly) with 274 members serving five-year terms; Zanzibar House of Representatives with 50 members serving five-year terms
Court system:	Court of Appeal
Armed forces:	27,000 troops
National holiday:	Union Day - April 26 (1964)
Estimated 2010 population:	43,526,000
Population density:	119 persons per mi² (46 per km²)
Population distribution:	75% rural, 25% urban
Life expectancy in years:	Male, 50; female, 53
Doctors per 1,000 people:	Less than 0.05
Birth rate per 1,000:	38
Death rate per 1,000:	13
Infant mortality:	73 deaths per 1,000 live births
Age structure:	0-14: 44%; 15-64: 53%; 65 and over: 3%
Internet users per 100 people:	1.2
Internet code:	.tz
Languages spoken:	Swahili or Kishwahili (official), English (official), Arabic, many local languages
Religions:	Mainland: Muslim 35%, indigenous beliefs 35%, Christian 30% Zanzibar: Muslim over 99%
Currency:	Tanzanian shilling
Gross domestic product (GDP) in 2008:	$20.61 billion U.S.
Real annual growth rate (2008):	7.1%
GDP per capita (2008):	$507 U.S.
Goods exported:	Cashew nuts, coffee, cotton, fish, gold, tobacco
Goods imported:	Chemicals, iron and steel, machinery, petroleum products, transportation equipment
Trading partners:	China, India, South Africa, Switzerland, United Arab Emirates

Mount Kilimanjaro, a dormant volcano in northern Tanzania, is Africa's highest mountain. The peak is surrounded by wildlife parks in Kenya and Tanzania.

zibar, the nation of Tanzania includes Pemba and several smaller islands. Zanzibar is the largest coral island off the African coast. The mainland coast and the islands are the hottest, wettest parts of Tanzania.

Plateaus rise gradually from the coastal lowlands. The Maasai Steppe in northeastern Tanzania lies about 3,500 feet (1,100 meters) above sea level, and west of the steppe the grassy central plateau reaches about 4,000 feet (1,200 meters). These plateaus, which are generally drier than the rest of the country, are covered mostly by grasses or barren land, with scattered patches of trees and shrubs.

Highlands rise in several areas of Tanzania. The north has some of the country's highest peaks, including Kilimanjaro, the highest point in Africa. Other highland areas rise in the central and southern regions. The highlands are generally warm and wet, but Kilimanjaro is so high that it remains snow-capped the year around.

The Rufiji, which flows east from the southern highlands, is Tanzania's main river. Part of Lake Victoria, the largest lake in Africa, lies in northern Tanzania. Lakes Tanganyika and Nyasa lie along the country's western border.

Tanzania lies on the eastern coast of mainland Africa and includes several nearby islands. Dar es Salaam is Tanzania's largest city, and Dodoma is the capital.

PEOPLE AND ECONOMY

Tanzania, one of Africa's poorest countries, is also one of the continent's most peaceful. Tanzania has not suffered the ethnic violence that has troubled so many other African nations.

Nevertheless, Tanzania has a great mix of different groups. About 98 percent of its people are of African ancestry. They belong to about 120 ethnic groups, including the Chagga, Maasai, Makonde, Nyamwezi, and Sukuma peoples. No single group is large enough to control the country, and this ethnic balance has helped the government develop a sense of national unity among Tanzanians. In addition to Africans, about 2 percent of Tanzania's people are descendants of Arabs, Europeans, Asian Indians, and Pakistanis.

Traditional clothing in Tanzania is a wrap-style garment for both men and women, but Western-style shirts and pants are increasingly popular with men. Muslim men often wear a flowing white robe called the kanzu.

Swahili and English are the official languages. Swahili—a blend of Arab and African languages—is more commonly used in everyday speech, and most Tanzanians also speak at least one tribal language.

About 75 percent of the people live in rural areas and farm for a living. Most farmers use old-fashioned tools, such as hoes and long-bladed knives, and grow only enough food to feed their families. The major food crops include bananas, cassava, corn, millet, rice, sorghum, wheat, and vegetables.

One of the most popular meals among Tanzanians is *ugali,* a porridge made with corn. The Maasai raise cattle on the interior plateaus, and cows' milk is an important part of their diet. Tanzanians who live along the coast or near large lakes also eat fish.

A Maasai market in northern Tanzania is a gathering place for rural people. Raising cattle is the chief activity of the Maasai and other peoples on the dry interior plateaus of Tanzania.

A coffee plantation benefits from the well-watered, fertile soil near the foot of Mount Meru. Rich farmland is found in these northern highlands, as well as in the southern highlands and the Lake Victoria region.

Severe economic problems in the 1980's changed some of these policies, however, and the government began to open more areas to private business. The nation's leaders have also tried to encourage industrial development, but most industries in the country remain small and unprofitable.

Manufacturing contributes about 9 percent of Tanzania's economic production. Factory workers process food and produce fertilizer, petroleum products, and textiles.

Service industries account for nearly 40 percent of the country's economic production. Service industries include banking, education, government, health care, insurance, and trade. Tourism—especially in the national parks—also aids Tanzania's developing economy.

Although only about 5 percent of Tanzania's land area is used for farming, agriculture employs about 80 percent of the country's work force. Most of the large, government-operated farms that produce many of Tanzania's export crops were private plantations under British rule. Much of Tanzania's export income comes from trading coffee, cotton, gold, tea, and tobacco.

When Tanzania was following strict socialism in the 1960's and 1970's, the government encouraged farmers to move from their small, scattered villages to larger *ujamaa* (cooperation) villages. Ujamaa villages were supposed to increase farm production and make it easier to provide the people with health and educational services. At one point during the 1970's, the government began forcing millions of people to move to the ujamaa villages.

A Zanzibar village is surrounded by lush vegetation. Zanzibar's exports include products of clove trees, one of the many species that flourish in the island's tropical heat and humidity.

SERENGETI NATIONAL PARK

The first Europeans to visit the Serengeti Plain were awe-struck by the landscape and wildlife they discovered there. Vistas of gently waving grass seemed to go on forever, and the abundance of wild game was unmatched by anything they had seen in their native lands.

The Serengeti is still known around the world for its vast and varied population of animals. Zebras, gazelles, and gnus, or wildebeests, travel in vast herds, while lions, cheetahs, and other predators hunt on the open plains. Antelopes, baboons, Cape buffaloes, elephants, flamingos, giraffes, hippopotamuses, monkeys, ostriches, rhinoceroses, and vultures are just a few more of the Serengeti's wild inhabitants.

The Serengeti National Park seeks to protect this wildlife. In the mid-1900's, Maasai herders were prohibited from grazing their cattle on the land, and rangers have since tried to protect the Serengeti wildlife from hunters. However, the plain and its wildlife do not stop at the park's borders—a fact that becomes more dramatically evident every year, when the great migration takes place.

Tourists observe a pride of lions from the safety of their vehicles.

Huge herds of gnus cross the Serengeti on their seasonal migration.

POACHING

To preserve its rich wildlife heritage, Tanzania, like many other African countries, has created animal preserves where the killing of animals is prohibited or limited. But poachers—people who hunt illegally—are a serious problem. Some poachers are Africans who are struggling to feed their families, but others hunt for profit. Elephant poachers, for example, kill elephants for the animals' ivory tusks. In 1976, the International Union for Conservation of Nature and Natural Resources, together with the World Wildlife Fund, established an elephant conservation organization. The sale of ivory has since been banned in numerous countries around the world, but the illegal slaughter of elephants and other animals continues.

A Verreaux's eagle clutches its prey in its talons as it soars above the rolling expanse of the Serengeti.

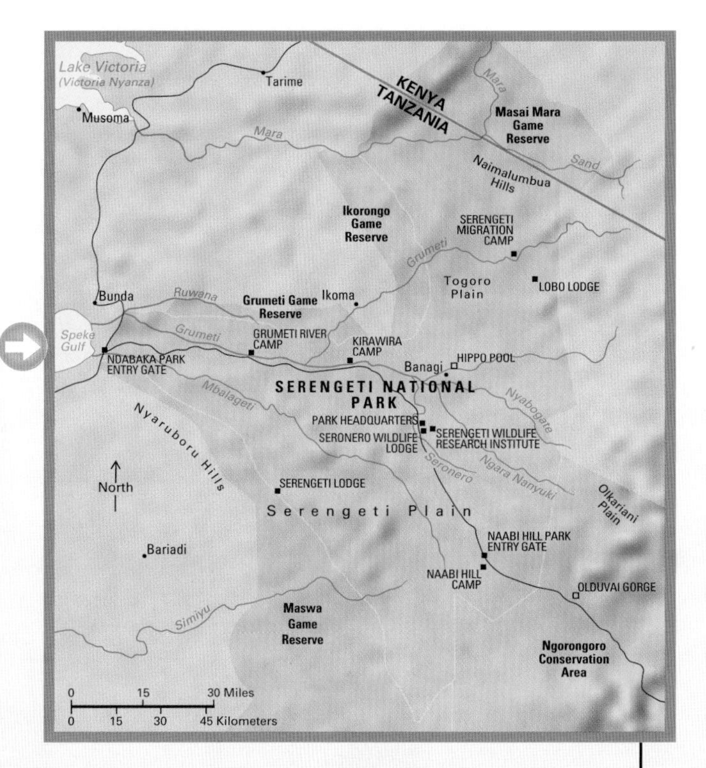

Serengeti National Park covers about 5,600 square miles (14,500 square kilometers) in northern Tanzania. The Serengeti Plain extends into neighboring Kenya as the Maasai Mara Game Reserve.

In the dry season, which lasts from late May to October, the plain cannot support many plants. So at the beginning of the dry season, huge herds of zebras, gnus, and gazelles leave the short-grass southeastern section of the Serengeti. They move northwest, near Lake Victoria, where rain is more plentiful, and trees and grasses provide them with food.

More than a million animals migrate across the long-grass savanna of the Serengeti to the woodland lake region. From there, they turn north and east to the Mara region in Kenya, before finally returning to the short-grass southeastern plain in November, at the end of the dry season.

The whole migration—a great triangular journey—covers about 500 miles (800 kilometers). Along the migration route, young or weak animals provide food for lions, leopards, and cheetahs. Animal carcasses provide food for the Serengeti's scavengers—the vultures, hyenas, and jackals. And also along the way, the migrating herds meet many nonmigrating animals, such as Cape buffaloes, hippopotamuses, and giraffes.

Serengeti National Park now includes about half of the entire Serengeti area. A scientific wildlife research center within the park's boundaries studies the area's fascinating ecology.

Giraffes abound on the grasslands of the Serengeti Plain. Lions are the only animals that attack adult giraffes, but young giraffes may be killed by other predators. Each giraffe has a spot pattern that distinguishes it from other individuals.

THAILAND

Thailand, in Southeast Asia, is a wet, tropical country with many rivers, forests, and mountains. The heartland of the country is the vast flood plain of the Chao Phraya River, which lies north of Bangkok, Thailand's capital. The river is an important means of transportation within Thailand, as well as a major source of irrigation for the many rice fields on the Central Plain.

Most of the people of Thailand—called *Thai*—live in villages ranging in size from several hundred to several thousand people. The houses are made of wood and thatch and are built on stilts to protect against flooding. In the wet season, families keep their farm animals under their homes, penned in by the stilts. High roofs keep the inside temperature relatively cool.

The majority of the Thai people are farmers, and rice is their leading crop. In the remote villages of the north, mountain streams deposit mud and sand along the banks during the rainy season and help make the soil fertile. Farmers grow rice in the narrow mountain valleys.

Rice and teak once provided Thailand with a large percentage of its export income. By the late 1980's, however, manufactured goods provided more than half of Thailand's foreign earnings. Tourism is also an important source of income. The magnificent buildings and historic monuments of Bangkok attract millions of visitors every year.

Since the 1960's, Bangkok and other Thai cities have drawn large numbers of people from rural areas—particularly young people seeking jobs and educational opportunities. The rapid population growth in Thailand's cities has resulted in high rates of unemployment and crowded living conditions. The government has initiated housing projects in an effort to ease the shortage.

Thailand has the distinction of being the only nation in Southeast Asia that has never been ruled by a Western power. The first Thai nation was established in A.D. 1238. The country was called Siam from 1782 until 1939—when it became known as Thailand. Its name in the Thai language is Muang Thai, which means Land of the Free.

THAILAND TODAY

Thailand's government is a constitutional monarchy. A legislature called the National Assembly consists of a House of Representatives and a Senate. The House selects a prime minister from among its members.

In local government, the people elect their own headman in each Thai village. The villages are grouped into units called *tambons*. The people in each tambon select from among the village headmen a *kamnan,* or chief administrator, for their tambon. Citizens at least 18 years of age must vote.

Shift from military rule

Thailand was under military rule from 1938 until 1973, when university students led a revolt against the military rulers. For the next three years, government officials were elected in a democratic fashion.

This period ended in October 1976, after conservative groups attacked radical students at Thammasat University in Bangkok. Many people were killed or arrested, and the military seized control of the government. From 1976 to 1988, the military gradually allowed a parliament to have more power.

In 1991, military leaders, charging the elected government with corruption, staged a peaceful coup. They removed the prime minister from office, dissolved the legislature, and appointed an interim (temporary) civilian government. A new constitution was adopted in late 1991, and a new civilian government was formed in March 1992. But in May, the prime minister was forced to resign amid protests. A new National Assembly was elected in September 1992, and elected civilian leaders took control of the government. A new constitution was adopted in 1997.

The 2000's

In 2001, a party called Thai Rak Thai (Thai Love Thai) won National Assembly elections, and party leader and telecommunications billionaire Thaksin Shinawatra became prime minister. Thaksin's government revived the economy, which had suffered a downturn in the late 1990's.

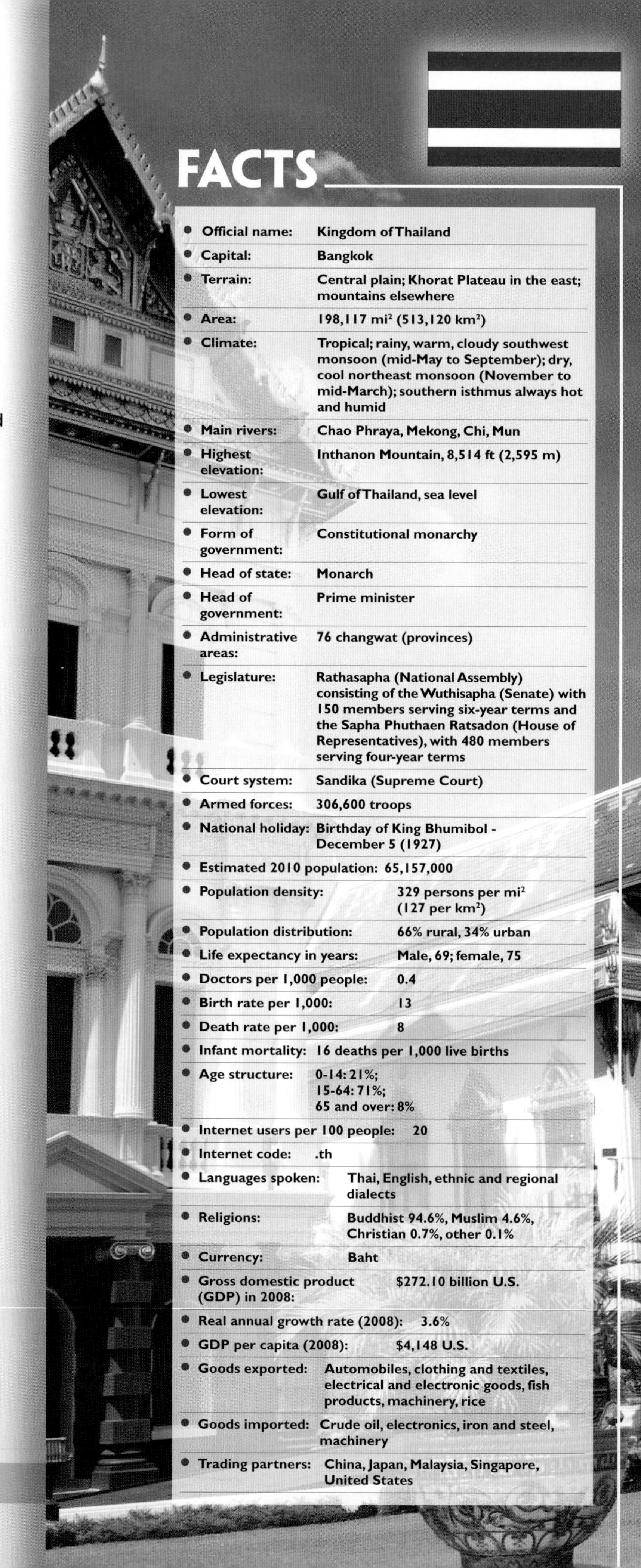

FACTS

Official name:	Kingdom of Thailand
Capital:	Bangkok
Terrain:	Central plain; Khorat Plateau in the east; mountains elsewhere
Area:	198,117 mi² (513,120 km²)
Climate:	Tropical; rainy, warm, cloudy southwest monsoon (mid-May to September); dry, cool northeast monsoon (November to mid-March); southern isthmus always hot and humid
Main rivers:	Chao Phraya, Mekong, Chi, Mun
Highest elevation:	Inthanon Mountain, 8,514 ft (2,595 m)
Lowest elevation:	Gulf of Thailand, sea level
Form of government:	Constitutional monarchy
Head of state:	Monarch
Head of government:	Prime minister
Administrative areas:	76 changwat (provinces)
Legislature:	Rathasapha (National Assembly) consisting of the Wuthisapha (Senate) with 150 members serving six-year terms and the Sapha Phuthaen Ratsadon (House of Representatives), with 480 members serving four-year terms
Court system:	Sandika (Supreme Court)
Armed forces:	306,600 troops
National holiday:	Birthday of King Bhumibol - December 5 (1927)
Estimated 2010 population:	65,157,000
Population density:	329 persons per mi² (127 per km²)
Population distribution:	66% rural, 34% urban
Life expectancy in years:	Male, 69; female, 75
Doctors per 1,000 people:	0.4
Birth rate per 1,000:	13
Death rate per 1,000:	8
Infant mortality:	16 deaths per 1,000 live births
Age structure:	0-14: 21%; 15-64: 71%; 65 and over: 8%
Internet users per 100 people:	20
Internet code:	.th
Languages spoken:	Thai, English, ethnic and regional dialects
Religions:	Buddhist 94.6%, Muslim 4.6%, Christian 0.7%, other 0.1%
Currency:	Baht
Gross domestic product (GDP) in 2008:	$272.10 billion U.S.
Real annual growth rate (2008):	3.6%
GDP per capita (2008):	$4,148 U.S.
Goods exported:	Automobiles, clothing and textiles, electrical and electronic goods, fish products, machinery, rice
Goods imported:	Crude oil, electronics, iron and steel, machinery
Trading partners:	China, Japan, Malaysia, Singapore, United States

In December 2004, a powerful undersea earthquake in the Indian Ocean generated a series of towering ocean waves called a *tsunami*. The tsunami killed about 228,000 people in Asia and Africa and left millions of others homeless. About 8,000 people along Thailand's southwestern coast were killed.

In 2006, the military overthrew Thaksin's government and appointed a temporary one. Voters approved a new constitution in 2007. In January 2008, a newly elected parliament chose the leader of the People's Power Party (PPP) as prime minister. However, he was soon removed from office for breaking conflict-of-interest rules, and a new PPP leader became prime minister. Protests against the PPP grew during the year. In December, the country's Constitutional Court dissolved the PPP, and parliament elected the Democrat Party leader as prime minister.

Anti-government protests began again in early 2009 and quickly became violent. Nearly 100 people were killed. Some of the protesters were arrested while others fled abroad. In 2011, Yingluck Shinawatra, Thaksin's youngest sister, led the Puea Thai Party to victory in general elections and became prime minister.

Thailand is bordered by Cambodia, Laos, Malaysia, and Myanmar. Southern Thailand extends along the Malay Peninsula.

Thailand's monarchy has done much to modernize Thailand. Thai kings, such as King Bhumibol Adulyadej, have built roads and railways, founded universities, and encouraged foreign travel. Only men could serve as monarch until 1980, when a new law decreed that either men or women could hold the office.

LAND AND PEOPLE

Thailand has four main land regions: the Mountainous North, the Khorat Plateau, the Central Plain, and the Southern Peninsula.

The Mountainous North extends along the country's western border to the Malay Peninsula. Thailand's highest peak, Inthanon Mountain, rises 8,514 feet (2,595 meters) in these mountains. The region has some forests of evergreen and teak.

Many streams and rivers flow south from the Mountainous North to the Gulf of Thailand, depositing rich silt along their banks. Farmers grow rice in the fertile soil of the mountain valleys, and the region also contains mineral deposits.

The Khorat Plateau, in northeastern Thailand, covers about 30 percent of the country's land area. The plateau is bordered by mountains on the south and west and by the Mekong River on the north and east. The Mekong, Chi, and Mun rivers provide irrigation for crops grown in the region.

The Central Plain, a richly fertile area between the Mountainous North and the Gulf of Thailand, is the nation's major rice-growing region. Four rivers unite in the northern part of the plain to form the Chao Phraya River, an important transportation route. The river empties into the Gulf of Thailand south of Bangkok.

The Southern Peninsula, which forms part of the Malay Peninsula, consists mainly of jungle with some mountains and rolling hills. Streams flow through the narrow valleys and flood small coastal plains. In the northern part of the peninsula, only a narrow strip of land belongs to Thailand, while the land in the western section belongs to Myanmar.

The southern part of the region occupies the entire width of the Malay Peninsula. Plantations of rubber trees thrive in this mountainous area. The area also contains rich deposits of tin.

The traditional dress of the Yao people features black turbans and trimmings of red wool. Government efforts to bring the mountain people into mainstream life have met with little success in Thailand.

Children play beside a dam on the Khorat Plateau. Such dams, built to confine the waters of the rivers that flow through the plateau, are an important part of the region's irrigation system. The sandy soil of this area holds little moisture.

Young girls of the Lisu ethnic group dress in traditional costumes to celebrate the New Year. The tribal peoples of the far north speak their own languages and follow many of their traditional customs.

Thailand's forests and jungles are home to a large variety of animals, including wild pigs, crocodiles, deer, tigers, and many types of poisonous snakes. At one time, elephants roamed wild across the land. Today, however, most of Thailand's elephants have been domesticated and trained to aid farmers and forestry workers.

Most of Thailand's people belong to the Thai ethnic group. The ancestors of the Thai migrated from southern China over a period of several centuries, beginning more than 1,000 years ago. At various times, other peoples also came to the area as migrants, war prisoners, or refugees. These peoples included Mons, Khmers, Indians, Vietnamese, and Malays. Most descendants of these groups now consider themselves Thai.

During the 1800's and 1900's, many Chinese came to Bangkok and other cities and towns. They often intermarried with the Thai and adopted local customs. Many of them now think of themselves as Thai of Chinese descent. Small, isolated groups, such as the Hmong, Karen, and Lua, live in the hills of the north and northwest.

About two-thirds of the Thai people live in rural villages. Most make their living by farming, though some work in the fishing and mining industries. The villagers grow most of their own food, including rice, corn, fruit, and cassava, a tropical plant whose roots are used in making tapioca.

Almost every village has a school and a *wat* (Buddhist temple). About 95 percent of the Thai people are Buddhists, and religion is an important part of everyday life in Thailand. According to custom, many Thai men become monks for at least a short time.

Since the 1960's, a huge number of rural people have migrated to urban areas, hoping to find work in industry and government. This rapid increase in city population has resulted in serious housing shortages. In Bangkok, for example, many people live in shacks in crowded slums.

Most Thai people wear the same clothing styles worn in Europe and North America. But in rural areas, some men and women wear the traditional *panung,* a colorful garment that wraps tightly around the body. A man's panung extends from the hips to the ankles. A woman's panung extends from above the chest to the ankles.

A sculpture of a bull stands in Siam Paragon in Bangkok. Siam Paragon, which opened in 2005, is one of the largest shopping centers in Asia.

ECONOMY

Before 1960, Thailand's economy relied almost entirely on the income received from rice, rubber, tin, and teak. In the 1960's, however, the government launched programs to stimulate the rural economy. Many new roads were constructed, vast forestlands were cleared, and banks loaned farmers money for agricultural improvements. Irrigation programs prolonged the growing season by providing storage dams, tractors replaced water buffaloes on farms, and better strains of rice produced more plentiful harvests.

Western European nations provided much of the funding needed for these improvements. The United States also paid Thailand large sums for permission to build U.S. air bases in the country during the Vietnam War (1957-1975).

After the mid-1980's, Thailand's economy also benefited from economic changes in Japan, Taiwan, and South Korea. The changes made it less profitable for manufacturing businesses to operate in those countries, and so many manufacturers moved their facilities to Thailand. The businesses were drawn to Thailand in part because land and labor were inexpensive. Manufacturing has become increasingly important in Thailand since the late 1900's.

Government, education, trade, transportation, and other service industries employ many people in Thailand. Other people work in construction and mining. The gap between the rich and the poor in Thailand is among the largest in the world.

Farm workers gather the rice harvest on Thailand's Central Plain. Rice is one of the country's chief exports, along with automobiles, electronic goods, fish products, and textiles.

Agriculture, forestry, and fishing

About 40 percent of Thailand's workers make their living by farming, and most farmers own their land. However, farming and fishing make up only about 10 percent of the nation's economy. Thai farmers mostly grow rice. Other major crops include cassava, coconut, mangoes, natural rubber, palm oil, pineapples, and sugar cane.

Forest products, especially teak, were once an important source of income in Thailand. In 1961, forests covered about 60 percent of the country. But exploitation of this resource and the clearing of forestland for agriculture reduced the

ELEPHANTS AT WORK

Trained elephants played an important role in Thailand's economy during the heyday of the logging industry. Forestry workers depended on the elephants' great strength to haul teak logs to rivers, where the logs were then floated downstream to sawmills. Thailand's government has banned logging within Thailand, but the log-processing industry continues to operate with teak and other wood imported from Myanmar. Elephants are trained to perform their work at an early age by their *mahout* (keeper). The mahout spends many hours with the elephant, perfecting the animal's skill and earning its trust. An annual elephant roundup, held in early November, allows the mahouts to show the public what their elephants can do.

A baby elephant drags a large teak log from a forest clearing under the supervision of its mahout. Before an elephant can be put to work, it must be trained through many mounting, marching, and log-dragging drills.

forests to about 30 percent by 1985. In 1988, Thailand announced that it would ban all logging within the country. Thai companies then arranged to cut teak and other tree species in Myanmar for processing in Thailand.

The rivers and coastal waters of Thailand provide plentiful supplies of anchovies, mackerel, and tuna. Large numbers of Thai people fish for a living. Many farmers also maintain well-stocked fishponds.

Palm trees fringe a peaceful beach on the Gulf of Thailand. Hotels and recreational facilities are being developed along the beaches to accommodate visitors to the region's beautiful and unspoiled coastline.

Manufacturing, mining, and tourism

Today, manufacturing and service industries, such as tourism, make up about 45 percent of the country's gross domestic product—that is, the value of all goods and services produced in the country in a year. However, only about 15 percent of the Thai people work in manufacturing industries. About 30 percent work in service industries.

Thailand's leading industries produce cement, food products, and textiles, particularly silk. In the Bangkok area, many international companies operate factories that assemble automobiles and electronic equipment and manufacture drugs and other products.

Thailand has valuable mineral deposits. The nation is among the world's leading producers of tin. Mines throughout the country produce barite, feldspar, gypsum, kaolin, lead, lignite, limestone, rock salt, and zinc, as well as precious stones. In addition, natural gas deposits were discovered in the Gulf of Thailand in the early 1980's.

Tourism is another important source of income for the people of Thailand. Each year, millions of people visit the nation's historic sites, relax at its coastal resorts, or simply enjoy the lively atmosphere of Bangkok's floating markets, roadside food stalls, and colorful night life.

Extracting fiber from the cocoons of silkworms, by means of a heating process, is one of the first steps in the production of silk. The raw silk is later woven into fabric. Embroidered silk fabrics are among Thailand's most popular products.

Workers extract salt from a lake in Thailand. Thailand's mineral wealth includes tin, precious stones, and other minerals, as well as deposits of natural gas and oil.

BANGKOK

Bangkok, Thailand's capital and largest city, is also the nation's chief commercial, cultural, and industrial center. The Thai name for Bangkok is *Krung Thep,* which means *City of Angels* or *Heavenly City.* Most of Bangkok lies on the east bank of the Chao Phraya River, about 17 miles (27 kilometers) north of the Gulf of Thailand, but the city also includes an area on the west bank called Thon Buri.

Bangkok is an extremely crowded city. Day and night, the streets are jammed with buses, bicycles, cars, and motorbikes. Some people prefer to weave through the traffic on a *tuk-tuk*—a small, three-wheeled, motorized cart.

Bangkok was once called the "Venice of the East" because of its many canals. By 1990, however, most of the canals had been filled in and replaced by streets and highways. Water taxis run on Bangkok's remaining canals and also on the Chao Phraya River. Bangkok's finest temples and palaces are located near the river.

Gilded spires and Buddhist images grace the roof of the Grand Palace. Built in the late 1700's as the royal residence of the Thai kings, the palace is now used mainly for ceremonial occasions.

Boats filled with fruits and vegetables line a canal at Damnoen Saduak, a floating market south of Bangkok. The city's remaining canals connect Bangkok's smaller, outlying communities with the city center.

Tourist attractions

The city contains more than 400 wats (Buddhist temples). One of the most famous is the Temple of the Emerald Buddha, which lies within the grounds of the Grand Palace, overlooking the Chao Phraya River. Once the home of the Thai kings, the palace is used mainly for state ceremonies today. The royal family lives in the Chitlada Palace, about 1 1/4 miles (2 kilometers) northeast of the Grand Palace. Bangkok's government buildings and cultural center lie between the two palaces.

The city's main business and commercial district is about 3 miles (5 kilometers) east of the Grand Palace. Bangkok's broad, modern avenues are lined with high-rise office buildings and department stores, fine shops, nightclubs, and motion-picture theaters.

Alongside these modern buildings, traditional Bangkok life goes on, with its colorful street vendors and its endless bustling markets. Rows of food stalls offer a wide variety of snacks, and the tempting aroma of herbs and spices fills the air. Even more foods can be found in the department stores, where Thai specialties fill entire floors.

Dancers in elaborately embroidered costume perform the stately movements of a traditional *lakhon* dance. Originally created to entertain Thai royalty, these dances are now performed for the Thai people in Bangkok's theaters.

Bangkok's most colorful food displays, however, can be found on the city's waterways. At the floating market in Thon Buri, for example, vendors in small boats laden with fruits and vegetables vie for space in a crowded canal.

Present and past

Bangkok is a fast-growing metropolis. Today, more than 6 million people crowd the city. In the suburbs, single-family dwellings line the streets, but within the city, housing is a problem. Although most people live in wooden houses or in apartments above shops, many are forced to live in crowded and unsanitary slums.

Overpopulation and Bangkok's many factories have also created environmental problems in the city. Exhaust fumes from motor vehicles and factory emissions pollute the air, and the city's inadequate drainage system cannot cope with the heavy rains that strike Bangkok. Floods damage the city each year.

Bumper-to-bumper traffic jams characterize Bangkok's congested streets. However, drivers of the small and maneuverable tuk-tuks can often weave their way through the heavy traffic.

HISTORY

People have lived in what is now Thailand for tens of thousands of years. The Mon and the Khmer peoples had settled much of Southeast Asia more than 2,000 years ago.

The ancestors of most present-day Thai probably migrated from southern China over a period of many centuries, beginning between about the A.D. 400's and 700's. They established many settlements in the region. In 1238, they formed the first Thai nation and named it *Sukothai,* which means *Dawn of Happiness.* King Ramkhamhaeng, one of the early rulers of Sukothai, may have developed a writing system for the Thai language. Over the next hundred years, Sukothai expanded until it occupied most of present-day Thailand.

By the late 1300's, however, the nation had declined in importance. The kingdom of Ayutthaya, established in 1351, absorbed Sukothai. The city of Ayutthaya served as the Thai capital from the mid-1300's to the mid-1700's. During that period, Ayutthaya flourished both economically and culturally, and military conquests expanded the kingdom. Wars were waged against the Malays to the south, the Burmese from what is now Myanmar in the west, and the Khmer people of Cambodia to the east. In 1431, Thai forces captured the Cambodian capital of Angkor.

Dancers at the Sukothai Candle Festival celebrate the formation of the first Thai nation in 1238. Sukothai was the name given to both the kingdom and its capital city.

Foreign invasions

The first Europeans to come to Ayutthaya were Portuguese traders in the early 1500's. During the 1600's, Spain, England, France, Japan, and the Netherlands also established trade there. In the years that followed, many young Thai learned Western ideas and customs.

Meanwhile, European powers had begun to interfere in Ayutthaya's internal affairs, angering the Thai. As a result, all Europeans were forced to leave the country, all treaties with European nations were canceled, and contact with Westerners was forbidden.

In 1767, the Burmese invaded Ayutthaya and destroyed the capital. Soon afterward, however, Thai forces drove out the Burmese, and a new capital was established at Thon Buri. In 1782, General Phyra Chakri became ruler of the kingdom as Rama I. The capital was moved across the Chao Phraya River to Bangkok.

Cordial relations with the nations of Western Europe were restored during the rule of King Mongkut (Rama IV) in the 1850's and 1860's. He employed Western advisers and encouraged the study of Western languages and modern science. Mongkut's son, King Chulalongkorn (Rama V), continued his father's policies. During his reign, from 1873 to 1910, Chulalongkorn abolished slavery, reorganized the government, and established a public education system for all the nation's children.

Damaged automobiles litter a street on the resort island of Phuket after the area was devasted by a tsunami in 2004. The tsunami killed about 8,000 people in Thailand.

Wat Phra Singh, a monastery in Chiang Mai, was founded in 1345. It contains a bronze Buddha image dating from the 1400's or earlier. Chiang Mai was founded in the northern part of the country in 1296. It rivaled the capital city of Ayutthaya in splendor for many years. In the 1700's, Chiang Mai was destroyed by invading Burmese forces, but it was restored in the 1790's. Today, it is one of Thailand's largest cities.

A line of sculpted Buddhas links modern Thailand to the earliest period in its history. It is believed that Buddhism in Thailand may date back to the 200's B.C., when India sent missionaries to the region.

The modern period

In 1932, a group of French-educated Thai led a revolt against King Prajadhipok (Rama VII) and forced the king to change the government from an absolute monarchy to a constitutional monarchy. Three years later, Prajadhipok abdicated, yielding the throne to his 10-year-old nephew. A *regency council* (a group of temporary rulers) was established to govern the country for the young king, but in 1938 military officers within the government took control. In 1939, they changed the country's name to Thailand.

During World War II (1939-1945), Japan invaded Thailand. Eventually, the Thai government decided to cooperate, and Japan took over Thailand's harbors, airports, and railroads. However, a resistance group within Thailand worked against Japan throughout the war.

Military leaders governed Thailand from the end of World War II until 1973, when a student rebellion forced the military authorities to allow democratic elections. Since that time, power in Thailand has changed hands both as a result of military coups and democratic elections.

In 1967, in an effort to maintain peace and stability in the region, Thailand, Indonesia, Malaysia, the Philippines, and Singapore formed the Association of Southeast Asian Nations (ASEAN). ASEAN is not a defense organization, but it works to promote economic, cultural, and social cooperation among its member nations.

BUDDHISM IN THAILAND

About 95 percent of the Thai people are Buddhists. Most follow the Theravada school, the only one of the early schools of Buddhism that has survived. Theravada means Way of the Elders.

The Theravadans emphasize the importance of Buddha as a historical figure. Most scholars believe that Buddha lived in northern India during the 500's or 400's B.C. His real name was Siddhartha Gautama. Buddha is a title meaning Enlightened One.

Buddha taught his followers that people who wanted to achieve enlightenment had to free themselves of all desires and worldly attachments. However, he did not preach extreme self-denial. He believed that people should follow the Middle Way, a way of life that avoids the extremes of self-denial or self-indulgence.

Buddha also taught that existence was a continuing cycle of death and rebirth. Each person's position and happiness in life was determined by the individual's behavior in past lives. The ultimate goal was nirvana, a state of peace and happiness.

The Theravadans also stress the virtues of the monastic life. Monks live in chastity and self-imposed poverty. They dedicate themselves to study and meditation. The discipline of such a life is considered essential for those who wish to achieve nirvana.

A graceful golden spire, called a chedi, soars over other buildings in the compound of a wat (Buddhist temple). The nation's wats rank among the finest examples of Thai architecture.

The monastic life

Monks in their orange robes are a common sight in Thailand. Every day, they accept offerings of food from the faithful who gather at the side of a road or waterway. These offerings not only nourish the monks, but they also give the faithful an opportunity to earn spiritual merit—a reward of good fortune that will be enjoyed either in this life or in a future life.

Buddhist monks line up to receive rice from a Thai farmer. The bowls they carry are among their few possessions. Monks are expected to reflect the Buddhist ideal of detachment from all worldly distractions.

Buddhists within the *sangha,* or monastic order, are governed by ancient rules of conduct that affect every aspect of a monk's life. However, if a monk no longer wishes to follow the monastic way of life, he may be released from his duties and leave the monastery at any time.

Many Thai men become monks for at least a short time. Many Buddhist men choose to become monks during Thailand's rainy summer season. During this period, which lasts from June to October, rice farmers have less work to do, and by living as a monk for just a few months, a man may expect to gain merit. He will also earn increased respect in his community.

Even young boys may become novice monks for a period of time. Novices learn to read and write, and they study Buddhist texts. Many Thai families send their sons to the monastery so that the boys will receive a good education.

Buddhist women in Thailand may pursue a religious life as nuns. They live near the temple, shave their heads once a month as monks do, and wear plain white robes. They are also expected to adhere to the strict discipline of the monastic life.

Life for a novice monk includes hard work as well as prayer and meditation. Parents who cannot afford the tuition for Thailand's privately owned high schools often send their sons to a monastery for an education.

Animism

When the Thai people became followers of Theravada Buddhism, they did not abandon all of their earlier beliefs. Instead, they combined traditional Buddhist practice with animism, a belief in spirits that exist outside the body. The Thai believe that these spirits reside in trees and rocks and watch people's behavior very closely.

The Thai people believe that both good and evil spirits inhabit the world and that people can please the good spirits and gain good fortune by offering them food and incense. To protect themselves from evil spirits, many Thai wear clay or stone charms engraved with an image of the Buddha. Rural people often mark their skin with tattoos to scare the spirits away.

Golden images of Buddha line the walls of a temple. The temple is dominated by a huge stone figure of Buddha. As a mark of respect, Buddha's images are elevated above a worshiper's head.

TOGO

Togo is a small country in western Africa. It lies on the Gulf of Guinea, an arm of the Atlantic Ocean. The name *Togo* means *behind the sea* in Ewe, one of the most widely spoken languages in the country.

Almost all the people of Togo are native Africans, and about one-third of them practice traditional African religions. A majority of the Togolese live in rural areas, and most farm for a living. Food crops such as cassava, corn, millet, sorghum, and yams are grown on farms owned mainly by families. Cash crops such as coffee and cocoa are grown mainly in the south. But good farmland is scarce in Togo. Harvests are small, and income is low.

The Togo Mountains cross the nation from southwest to northeast, and they divide Togo into two major regions—the north and the south. The Togolese today share some common characteristics, but the many differences in their ways of life reflect their history.

The people in northern Togo live in villages of adobe houses with cone-shaped thatched roofs. Their ancestors were people who came from the west African *savanna* (thinly wooded grassland) region. The traditional dress of the people of northern Togo is a white cotton smock. Most of Togo's Muslims live in the north. Many different languages are spoken in the region.

The people in southern Togo, whose ancestors came from Benin and Ghana, live in *compounds,* or groups of huts enclosed by walls. The south has been more deeply influenced by European customs than the north has. Many southerners still wear traditional clothing—a loose-fitting toga—but others wear Western-style clothes. Some work for the government or run small businesses. The people of southern Togo are mainly Christian. Most speak the Ewe or Mina languages.

Historians believe that the people who live in the central mountains are the descendants of the original inhabitants of Togo. By the 1300's, Ewe-speaking people began to move into what is now southern Togo. They were followed by invaders from the north

FACTS

Official name:	**République Togolaise (Togolese Republic)**
Capital:	**Lomé**
Terrain:	**Gently rolling savanna in north; central hills; southern plateau; low coastal plain with extensive lagoons and marshes**
Area:	**21,925 mi² (56,785 km²)**
Climate:	**Tropical; hot, humid in south; semiarid in north**
Main rivers:	**Mono, Oti**
Highest elevation:	**Mont Agou, 3,235 ft (986 m)**
Lowest elevation:	**Atlantic Ocean, sea level**
Form of government:	**Republic**
Head of state:	**President**
Head of government:	**Prime minister**
Administrative areas:	**5 regions**
Legislature:	**National Assembly with 81 members serving five-year terms**
Court system:	**Cour d'Appel (Court of Appeal), Cour Supreme (Supreme Court)**
Armed forces:	**8,600 troops**
National holiday:	**Independence Day - April 27 (1960)**
Estimated 2010 population:	**7,091,000**
Population density:	**323 persons per mi² (125 per km²)**
Population distribution:	**59% rural, 41% urban**
Life expectancy in years:	**Male, 56; female, 60**
Doctors per 1,000 people:	**Less than 0.05**
Birth rate per 1,000:	**37**
Death rate per 1,000:	**10**
Infant mortality:	**65 deaths per 1,000 live births**
Age structure:	**0-14: 43%; 15-64: 54%; 65 and over: 3%**
Internet users per 100 people:	**6**
Internet code:	**.tg**
Languages spoken:	**French (official), Ewe, Mina, Kabye (or Kabiye), Dagomba**
Religions:	**Christian 47.1%, animist 33%, Muslim 13.7%, other 6.2%**
Currency:	**Communaute Financiére Africaine franc**
Gross domestic product (GDP) in 2008:	**$2.89 billion U.S.**
Real annual growth rate (2008):	**0.8%**
GDP per capita (2008):	**$435 U.S.**
Goods exported:	**Cement, cocoa, coffee, cotton**
Goods imported:	**Food, machinery, petroleum products, transportation equipment**
Trading partners:	**Burkina Faso, China, France, Ghana, Netherlands**

and war refugees from neighboring countries who settled in Togo between the 1500's and 1800's.

Europeans arrived in the region in the late 1400's, when the Portuguese came to explore and trade. Though a small slave trade took place at Aneho, on Togo's coast, most slave trading in the region was conducted at ports in what are now Benin and Ghana.

German traders and missionaries arrived in Togo in the late 1800's, and by 1899 Togo was a German protectorate. After Germany lost World War I (1914-1918), the United Kingdom and France took control of Togo. British Togoland became part of the Gold Coast (now Ghana) after World War II (1939-1945).

An independence movement arose in French Togoland, and on April 27, 1960, French Togoland became the independent Republic of Togo. Sylvanus Olympio, the leader of the independence party, became president. However, a rivalry between northerners and southerners led to violence. In 1963, a group of northern army officers assassinated Olympio, a southerner, and named Nicolas Grunitzky president.

In 1967, army officers led by Gnassingbé Eyadéma overthrew Grunitzky and set up a government with Eyadéma as president. The people of Togo endorsed Eyadéma in a 1972 vote and reelected him in 1979 and 1986. Multiparty elections held in 1993, 1998, and 2003 confirmed Eyadéma as president.

Eyadéma died in February 2005. The military installed his son, Faure Gnassingbé as president. Faced with international criticism, Gnassingbé stepped down but then won election as president in April 2005. He was reelected in 2010.

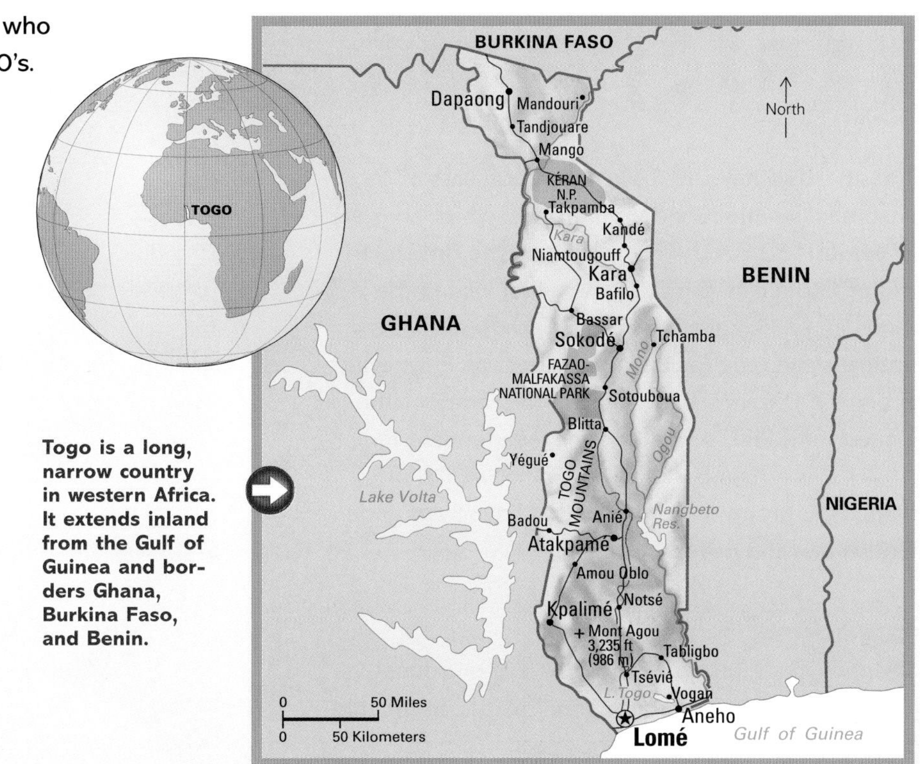

Togo is a long, narrow country in western Africa. It extends inland from the Gulf of Guinea and borders Ghana, Burkina Faso, and Benin.

Lomé, the capital of Togo, is a popular tourist destination. The church in the background dates from the time of German colonial rule.

TONGA

About 1,950 miles (3,140 kilometers) east of Australia lies the Kingdom of Tonga, which consists of about 150 South Pacific islands. The British explorer Captain James Cook, who first visited the islands in 1773, called them the *Friendly Islands.* Tonga is the only remaining kingdom of Polynesia. From 1900 to 1970, it was a protectorate of the United Kingdom. Tonga became independent in 1970. It is a member of the Commonwealth of Nations, a group of countries that have lived under British law and government.

Environment

Tonga is made up of three major island groups—Tongatapu, at the southern end; Ha'apai, in the middle; and Vava'u, at the northern end. Most of Tonga's people live on these islands, which are mainly coral reefs. A chain of higher, volcanic islands with some active volcanoes lies west of the coral islands. Nuku'alofa, the country's capital, chief port, and commercial center, is on Tongatapu, the largest island in Tonga.

At one time, tropical rain forests covered the islands. However, much of this forest area has been replaced by large plantations. Fertile clay soils cover most of Tonga, but deposits of sandy soil lie along the coasts.

Tonga has a warm, wet climate with high humidity. Temperatures average 78° F (26° C). The average annual rainfall in the kingdom varies from 70 inches (180 centimeters) on Tongatapu to 100 inches (250 centimeters) on some northern islands. Most rain falls from December through March. Cyclones—with their lashing rains, violent thunder, and lightning—sometimes strike Tonga. These storms may bring winds of up to 180 miles (290 kilometers) an hour. They may measure from 200 to 300 miles (320 to 480 kilometers) across.

Government

Tonga is a constitutional monarchy. The king is assisted by the Cabinet, which includes a prime minister and the governors of Ha'apai and Vava'u. The

FACTS

Official name:	Kingdom of Tonga
Capital:	Nuku'alofa
Terrain:	Most islands have limestone base formed from uplifted coral formation; others have limestone overlying volcanic base
Area:	289 mi² (748 km²)
Climate:	Tropical, modified by trade winds; warm season (December to May), cool season (May to December)
Main rivers:	N/A
Highest elevation:	Kao, an extinct volcano in the Ha'apai Group, 3,380 ft (1,030 m)
Lowest elevation:	Pacific Ocean, sea level
Form of government:	Constitutional monarchy
Head of state:	Monarch
Head of government:	Prime minister
Administrative areas:	3 island groups
Legislature:	Fale Alea (Legislative Assembly) with 32 members serving three-year terms
Court system:	Supreme Court, Court of Appeal
Armed forces:	N/A
National holiday:	Emancipation Day - June 4 (1970)
Estimated 2010 population:	104,000
Population density:	360 persons per mi² (139 per km²)
Population distribution:	75% rural, 25% urban
Life expectancy in years:	Male, 69; female, 73
Doctors per 1,000 people:	0.3
Birth rate per 1,000:	23
Death rate per 1,000:	5
Infant mortality:	12 deaths per 1,000 live births
Age structure:	0-14: 34%; 15-64: 61%; 65 and over: 5%
Internet users per 100 people:	8
Internet code:	.to
Languages spoken:	Tongan, English
Religion:	Christian
Currency:	Pa'anga
Gross domestic product (GDP) in 2008:	$260 million U.S.
Real annual growth rate (2008):	1.2%
GDP per capita (2008):	$2,430 U.S.
Goods exported:	Fish, vegetables and other agricultural products
Goods imported:	Machinery, meat, petroleum products, transportation equipment
Trading partners:	Australia, Fiji, Japan, New Zealand, United States

George V was crowned king of Tonga in 2008. He replaced his father, King Tupou IV, who died in 2006 after ruling the country for 40 years.

king appoints the prime minister and the two governors. Other Cabinet members are either elected or appointed. The Legislative Assembly is composed of the Cabinet, elected nobles, and elected commoners. Elections take place every three years. Tongans must be at least 21 years old and able to read and write in order to vote.

The Legislative Assembly meets for two or three months a year. When the Assembly is not in session, the Privy Council has the power to make laws. The Privy Council consists of the king and the Cabinet. However, the Legislative Assembly may change laws passed by the Privy Council.

The government publishes a weekly newspaper in both Tongan and English. It operates the country's only radio station.

Since independence, Tonga—with British aid—has worked to modernize its agriculture and to build wharves and airstrips. It has also encouraged foreign investment. Through these efforts, the government has sought to provide increased job opportunities for Tonga's growing population.

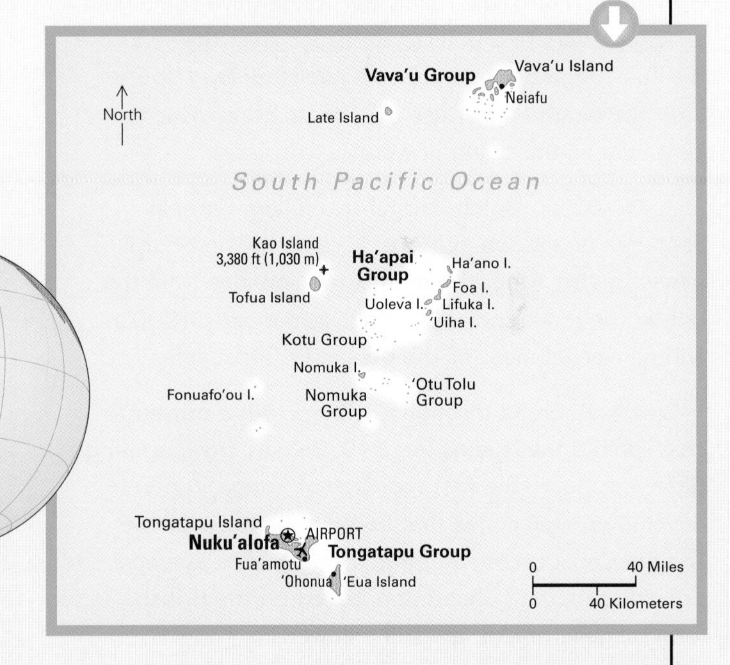

The Kingdom of Tonga lies in the South Pacific and consists of about 150 islands. The three major island groups, which stretch from north to south, are Vava'u, Ha'apai, and Tongatapu. They are formed mainly of coral reefs.

After the death of King Tupou IV in 2006, a government committee recommended that the Legislative Assembly be elected rather than appointed by the king. In November, riots triggered by prodemocracy demonstrations resulted in six deaths. Tonga's government then passed legislation requiring most members of parliament to be elected by the people. Tonga held its first democratic election in 2010.

PEOPLE

Tonga has a population of more than 100,000 people. Almost all Tongans are Polynesians and Christians. Most Tongans live in small villages and raise crops. The people also fish for such seafood as shark and tuna. Most of the islands have no running water, and many have no electric service. About two-thirds of Tonga's people live on Tongatapu, the largest island.

Settlers, early rulers, and independence

The first people to settle in Tonga were Polynesians who probably came from Samoa. Although much of Tonga's early history is based on myths, records of Tongan rulers go back about 1,000 years. The people believed that the early rulers, who held the hereditary title of *Tu'i Tonga,* were sacred representatives of the Tongan gods.

In about 1470, the ruling Tu'i Tonga gave some governing powers to a nonsacred leader. Over the years, the Tu'i Tonga became only a figurehead. By 1865, after the death of the last Tu'i Tonga, the nonsacred king held all the ruling power.

In 1616, two Dutch navigators, Willem Cornelis Schouten and Jakob le Maire, became the first Europeans to visit Tonga. Methodist missionaries from the United Kingdom settled in Tonga in the early 1800's and converted most of the people to Christianity.

Civil war spread throughout Tonga, but a powerful chief united the islands in 1845. He was crowned King George Tupou I, the first monarch of Tonga. Tupou I developed legal codes that became the basis of the Tongan Constitution, adopted in 1875. The essential principles of the Constitution, based on the British form of government, are still followed.

Tupou I was succeeded by his great-grandson, Tupou II, in 1893. In 1900, under Tupou II's rule, Tonga became a protectorate of the United Kingdom. Queen Salote, who came to the throne in 1918 and ruled until 1965, worked to improve education and health in Tonga. In 1970, Tonga gained independence from the United Kingdom.

Tongan girls in their school uniforms form a colorful group. The law requires Tongan children from 6 to 14 years old to go to school. Tongans must be able to read and write in order to vote.

Dancers adorned with flowers sail across a calm lagoon while, in the stern of the raft, two musicians play conch shells. People have used conch shells as musical instruments for thousands of years.

Way of Life

The Methodist missionaries of the 1800's had a lasting effect on Tonga, and many of Tonga's people remain Methodists today. Tonga's Constitution prohibits work or recreation on Sunday, and this law is strictly observed.

Fertile soils and a warm climate have also shaped Tongan life, and agriculture is still the basis of Tonga's economy. Most Tongans live in small villages and raise crops, such as bananas, breadfruit, pumpkins, sweet potatoes, tapioca, and yams. The chief exports are fish, vegetables, and other agricultural products.

About three-fourths of Tonga's workers are farmers, but the government owns all the land. Every male who is 16 or over is entitled to rent a plot of land from the government. However, there is not enough land to legally go around, and many of those who are legally entitled do not actually get land.

Baskets of yams are displayed for sale at a Tongan market. The nation's fertile soil and warm climate have made agriculture the basis of its economy.

Waves crash against the shore of one of Tonga's approximately 150 islands. Tonga's people fish for such seafood as shark and tuna.

The government has helped establish small-scale manufacturing in Tonga, especially in Nuku'alofa. Tourism is also important to the economy. The most important airport operates at Fua'amotu on Tongatapu.

The country's official languages are Tongan and English. The law requires all Tongan children from 6 to 14 years old to attend school. The government operates almost all of the schools. Churches direct the rest. The University of the South Pacific has a branch in Nuku'alofa. Tongan schoolchildren enjoy many sports, especially rugby football.

Traditional woodcarvings are popular souvenirs at a Tongan market stand. Tourism has become increasingly important to Tonga's economy.

TRINIDAD AND TOBAGO

Trinidad and Tobago is a country consisting of two islands in the Caribbean Sea. Trinidad, the largest island in the Lesser Antilles, lies 7 miles (11 kilometers) east of Venezuela, and Tobago lies about 20 miles (32 kilometers) northeast of Trinidad. This island country has been independent since 1962.

Tropical forests and fertile flatland cover much of the island of Trinidad. A mountain range extends across the northern area, and hills rise in the central and southern sections. Tobago has wide, sandy beaches along the coast and a mountain ridge in the center of the island. Both Trinidad and Tobago have a warm, moist climate.

History

Arawak and Carib Indians were living on Trinidad in 1498, when Christopher Columbus claimed the island for Spain. The Spaniards set up a permanent settlement on Trinidad in 1592. Four years later, a British sea captain reported seeing Tobago, and the Dutch settled there in 1632.

Trinidad's population did not begin to grow rapidly until 1783, when Spain offered land grants in Trinidad to any Roman Catholics willing to develop the island's economy. Colonists from Spain and settlers of French descent from other Caribbean islands accepted the offer and established thriving sugar cane plantations on the island.

In 1797, the British captured Trinidad and ruled it for more than 150 years. The British took control of Tobago in 1814, after fighting France for possession of the island. In 1889, Trinidad and Tobago became one colony under British rule.

The Great Depression of the 1930's brought economic setbacks for the people of Trinidad and Tobago, and the people began to demand a greater voice in their government. After a gradual increase in self-government during the 1940's and 1950's, the people of Trinidad and Tobago won their independence in 1962.

FACTS

Official name:	Republic of Trinidad and Tobago
Capital:	Port-of-Spain
Terrain:	Mostly plains with some hills and low mountains
Area:	1,981 mi² (5,130 km²)
Climate:	Tropical; rainy season (June to December)
Main rivers:	Caroni, Ortoire
Highest elevation:	Mount Aripo, 3,085 ft (940 m)
Lowest elevation:	Caribbean Sea, sea level
Form of government:	Republic
Head of state:	President
Head of government:	Prime minister
Administrative areas:	9 regional corporations, 2 city corporations, 3 borough corporations, 1 ward
Legislature:	Parliament consisting of the Senate with 31 members serving a maximum term of five years and the House of Representatives with 41 members serving five-year terms
Court system:	Supreme Court
Armed forces:	4,100 troops
National holiday:	Independence Day - August 31 (1962)
Estimated 2010 population:	1,345,000
Population density:	679 persons per mi² (262 per km²)
Population distribution:	87% rural, 13% urban
Life expectancy in years:	Male, 67; female, 72
Doctors per 1,000 people:	0.8
Birth rate per 1,000:	14
Death rate per 1,000:	8
Infant mortality:	30 deaths per 1,000 live births
Age structure:	0-14: 21%; 15-64: 72%; 65 and over: 7%
Internet users per 100 people:	24
Internet code:	.tt
Languages spoken:	English (official), Trinidad English, Hindi, French, Spanish, Chinese
Religions:	Roman Catholic 26%, other Christian 31.6%, Hindu 22.5%, Muslim 5.8%, other 14.1%
Currency:	Trinidad and Tobago dollar
Gross domestic product (GDP) in 2008:	$24.29 billion U.S.
Real annual growth rate (2008):	3.5%
GDP per capita (2008):	$18,456 U.S.
Goods exported:	Ammonia, methanol, natural gas, petroleum and petroleum products, steel products
Goods imported:	Chemicals, crude oil, food, machinery, transportation equipment
Trading partners:	Brazil, Jamaica, United States, Venezuela

Sun, sea, sand, and solitude are the ideal recipe for relaxation on the island of Tobago. Less developed and more scenic than Trinidad, Tobago attracts many tourists.

Trinidad and Tobago is an independent island country near the northeast coast of South America. It is the home of calypso music, the limbo dance, and the famous annual Trinidad Carnival.

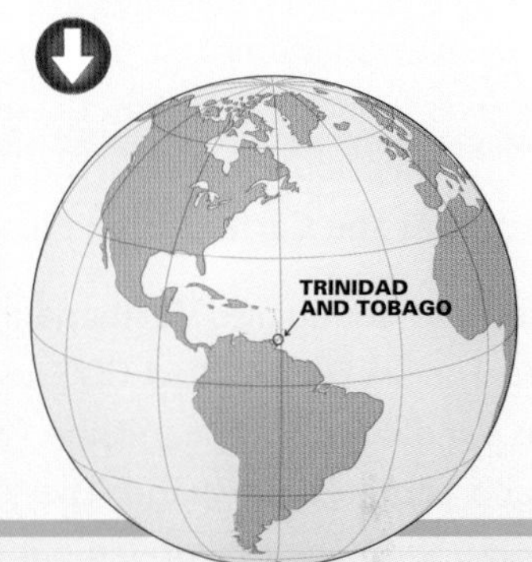

People and economy

About 40 percent of the people of Trinidad and Tobago are descended from African slaves who were brought to the islands to work on the plantations. Another 40 percent are descendants of workers from India who came to the islands after slavery was abolished in 1833. People of mixed European and African ancestry, plus groups of Europeans and Chinese, form the rest of the population.

English is the country's official language. Many people speak Trinidad English, a form of English with French and Spanish influences. French, Hindi, and Spanish are also spoken on the islands.

In the past, most of the people farmed for a living. However, the oil production and refining industry now employs many workers. Oil and gas products account for most of the country's export income and provide the basis for the nation's economy. Trinidad and Tobago is also one of the world's leading exporters of ammonia and methanol.

The towering smokestacks of an oil refinery dominate the bay of San Fernando, on the southwestern coast of Trinidad.

TUNISIA

Tunisia lies in the middle of northern Africa, on the Mediterranean Sea. It extends farther north than any other country in Africa. For centuries, people have entered northwest Africa through Tunisia.

About 3000 B.C., Berbers migrated into the region—probably from Asia or Europe—and settled there. Then, about 1100 B.C., the Phoenicians sailed to Tunisia from the eastern Mediterranean and established an empire. The powerful city of Carthage was founded near what is now the capital city of Tunis about 814 B.C. The empire came to be known as the Carthaginian Empire.

The Romans conquered Carthage in 146 B.C. and ruled Tunisia for the next 600 years. In A.D. 439, the Vandals, a European tribe, invaded Tunisia and captured Carthage. The Vandals ruled the region until they were driven out by the Byzantines of Constantinople in 534.

Like Morocco and Algeria to the west, Tunisia was invaded by Arab Muslims in the 600's. The Arabs called this entire northwest African region Maghreb—meaning *the place of the sunset—the west*. They destroyed Carthage and founded their own capital, Kairouan.

The Arab invasion was a turning point in Tunisia's history. Over time, the people began to speak the Arabic language and follow Arabic customs, and they adopted Islam.

The Ottoman Empire, which was centered in Asia Minor (now part of Turkey), took control of Tunisia in 1574. The Ottoman rulers appointed a *bey* (ruler) to govern Tunisia from Tunis, near the site of ancient Carthage.

By the 1700's, the Tunisian beys had achieved a large measure of independence in running the region's affairs. In 1881, France occupied Tunisia and made it a protectorate. The French controlled Tunisia's financial, foreign, and military affairs, leaving the beys with little authority. But Tunisia was technically a part of the Ottoman Empire until 1918, when the empire was on the losing side of World War I (1914-1918).

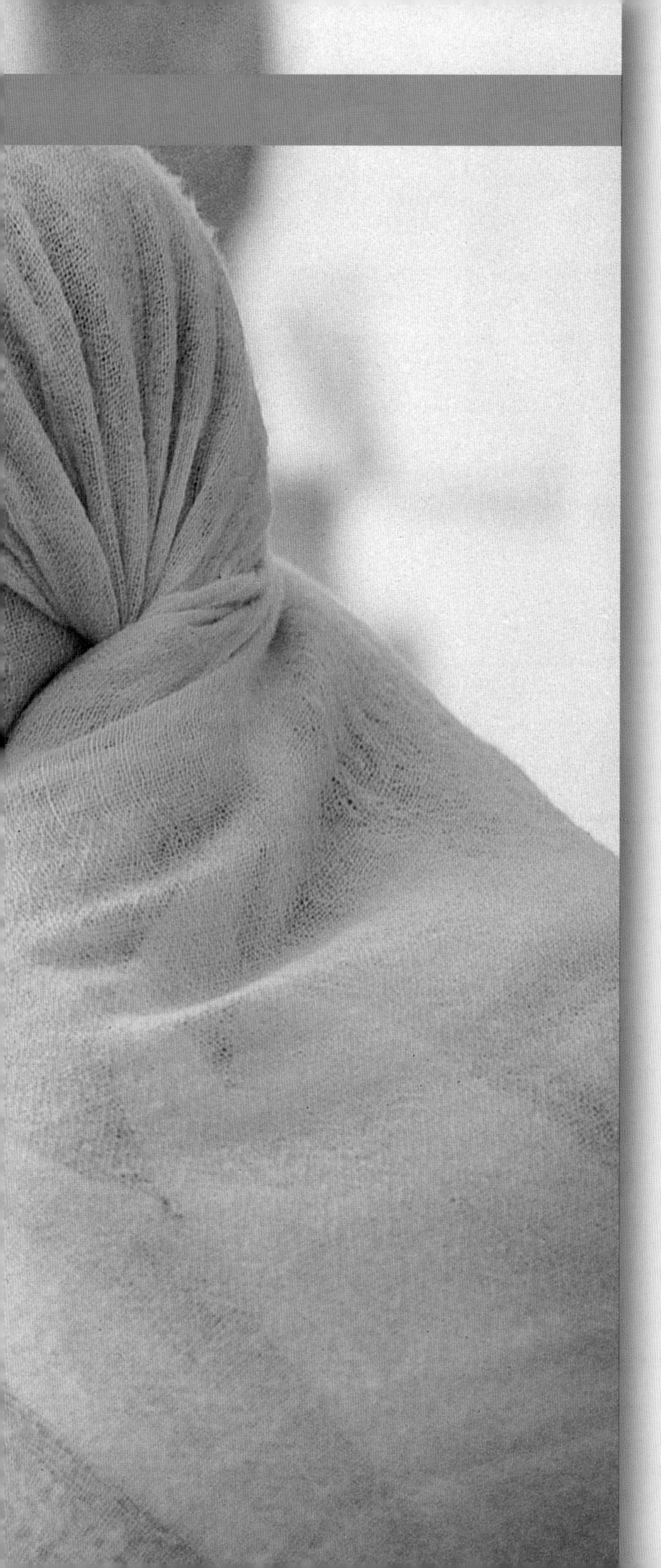

The first successful Tunisian independence movement began in 1934. That year, Habib Bourguiba founded a political party called the Neo-Destour (New Constitution) Party, now known as the Democratic Constitutional Rally Party. Bourguiba led the struggle for independence for more than 20 years.

In 1955, France finally allowed the Tunisians internal self-government. Full independence followed in 1956. However, France maintained troops and military bases in the country until the early 1960's.

Tunisia became a republic in 1957, and Bourguiba was elected president. He was reelected three times and in 1975, was named president for life. Bourguiba was extremely popular and respected by the Tunisian people. He introduced many economic and social reforms, including voting rights for women and the establishment of a national school system.

In the 1970's and 1980's, a growing number of Tunisians became discontented with single-party rule. In 1987, Prime Minister Zine el-Abidine Ben Ali removed Bourguiba from office, claiming that Bourguiba was no longer able to handle the responsibilities of the presidency. Ben Ali then became president. He promised new laws allowing more political parties and that he would serve only until the next election called for by Tunisia's constitution. He was elected president in 1989 and was reelected several times with little or no opposition.

Massive protests against Ben Ali's government began in late 2010. The protesters expressed anger over government corruption, a lack of accountability among the ruling family, and high unemployment, especially among youth. Similar protests soon erupted in several other countries of the Middle East. Ben Ali stepped down as president in January 2011 and fled the country. A new assembly was elected in October, and the assembly elected a government in December that was to serve until a new constitution could be written.

TUNISIA TODAY

Tunisia is part of the Arab world, the Mediterranean area, and Africa. Almost all Tunisians speak Arabic and follow an Arab way of life. But because France controlled Tunisia from 1881 until 1956, Tunisia also shows French influences. Tunis is Tunisia's capital and largest city.

Under Tunisia's original constitution, which was adopted in 1959 and revised in 1988 and 2002, Tunisia is a republic headed by a president. The president is elected by the people to a term of five years. The president appoints a Cabinet headed by a prime minister to help run the government. The president also appoints a governor to head each of Tunisia's 24 local *governorates.*

After Ben Ali fled the country in 2011, the people elected a new legislature called the Constituent Assembly. International observers praised the election as fair and democratic. Voters elected 217 members to the assembly, which then formed a caretaker government. The purpose of the assembly was to write a new constitution, which would pave the way for legislative and presidential elections.

Until a new constitution could be written and elections held, the assembly members agreed that three parties would divide the top governmental posts among themselves. Hamadi Jebali of the moderate Islamist Ennahda party, which won the most seats, became prime minister; Moncef Marzouki of the Congress for the Republic was chosen as president (a less powerful position than it had been under the previous Constitution); and Mustapha Ben Jaafar of the Ettakatol party became chair of the new assembly.

Two other parties, the Progressive Democratic Party and the Democratic Modernist Pole, formed the opposition. The Democratic Constitutional Rally Party—Tunisia's chief political party from 1957 until 2011—disbanded. Ben Ali and Bourguiba had both been members of the Rally party.

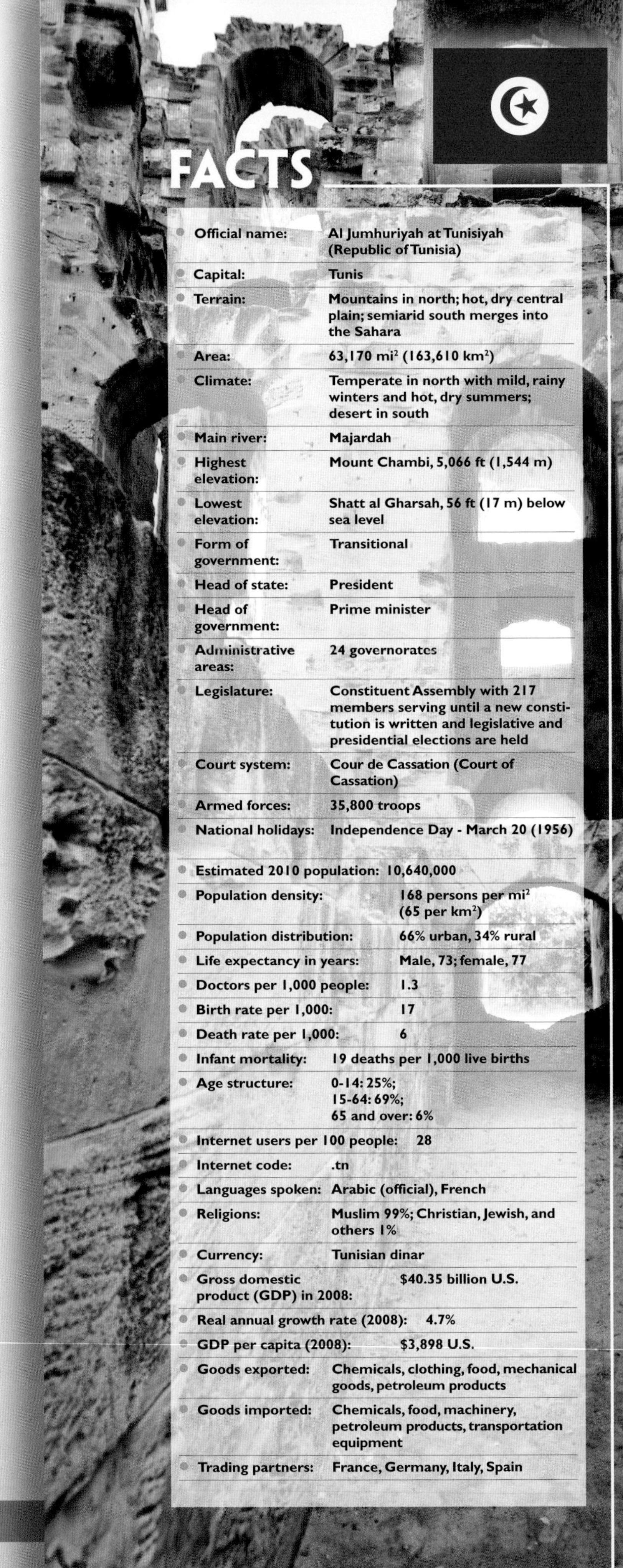

FACTS

Official name:	Al Jumhuriyah at Tunisiyah (Republic of Tunisia)
Capital:	Tunis
Terrain:	Mountains in north; hot, dry central plain; semiarid south merges into the Sahara
Area:	63,170 mi^2 (163,610 km^2)
Climate:	Temperate in north with mild, rainy winters and hot, dry summers; desert in south
Main river:	Majardah
Highest elevation:	Mount Chambi, 5,066 ft (1,544 m)
Lowest elevation:	Shatt al Gharsah, 56 ft (17 m) below sea level
Form of government:	Transitional
Head of state:	President
Head of government:	Prime minister
Administrative areas:	24 governorates
Legislature:	Constituent Assembly with 217 members serving until a new constitution is written and legislative and presidential elections are held
Court system:	Cour de Cassation (Court of Cassation)
Armed forces:	35,800 troops
National holidays:	Independence Day - March 20 (1956)

Estimated 2010 population:	10,640,000
Population density:	168 persons per mi^2 (65 per km^2)
Population distribution:	66% urban, 34% rural
Life expectancy in years:	Male, 73; female, 77
Doctors per 1,000 people:	1.3
Birth rate per 1,000:	17
Death rate per 1,000:	6
Infant mortality:	19 deaths per 1,000 live births
Age structure:	0-14: 25%; 15-64: 69%; 65 and over: 6%
Internet users per 100 people:	28
Internet code:	.tn
Languages spoken:	Arabic (official), French
Religions:	Muslim 99%; Christian, Jewish, and others 1%
Currency:	Tunisian dinar
Gross domestic product (GDP) in 2008:	$40.35 billion U.S.
Real annual growth rate (2008):	4.7%
GDP per capita (2008):	$3,898 U.S.
Goods exported:	Chemicals, clothing, food, mechanical goods, petroleum products
Goods imported:	Chemicals, food, machinery, petroleum products, transportation equipment
Trading partners:	France, Germany, Italy, Spain

Tunis, the capital and largest city of Tunisia, is the nation's political and cultural center.

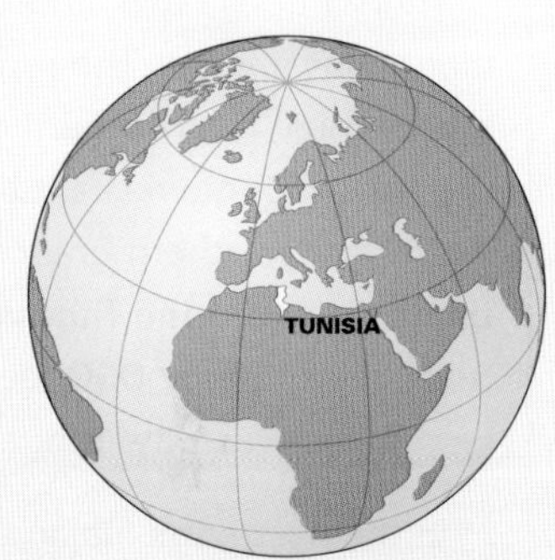

In December 2011, the Constituent Assembly adopted a provisional constitution, so that the government could continue to function. The so-called "mini-constitution" generated much heated debate over such issues as the role of Islam in the country and the power of the president.

As in most Arab countries, Islam had been the basis of the nation's traditional social and political values. But Ben Ali's government also made reforms that followed more liberal, Western ideas. The traditional religious courts, or *Sharī`ah* courts, were closed. The responsibility for caring for the elderly and people with disabilities was taken over by the government. Women were granted the right to vote, as well as near-equal status with men in such matters as property, divorce, and custody of children. Some Muslims were unhappy with such changes and favored a return to traditional Islamic practices.

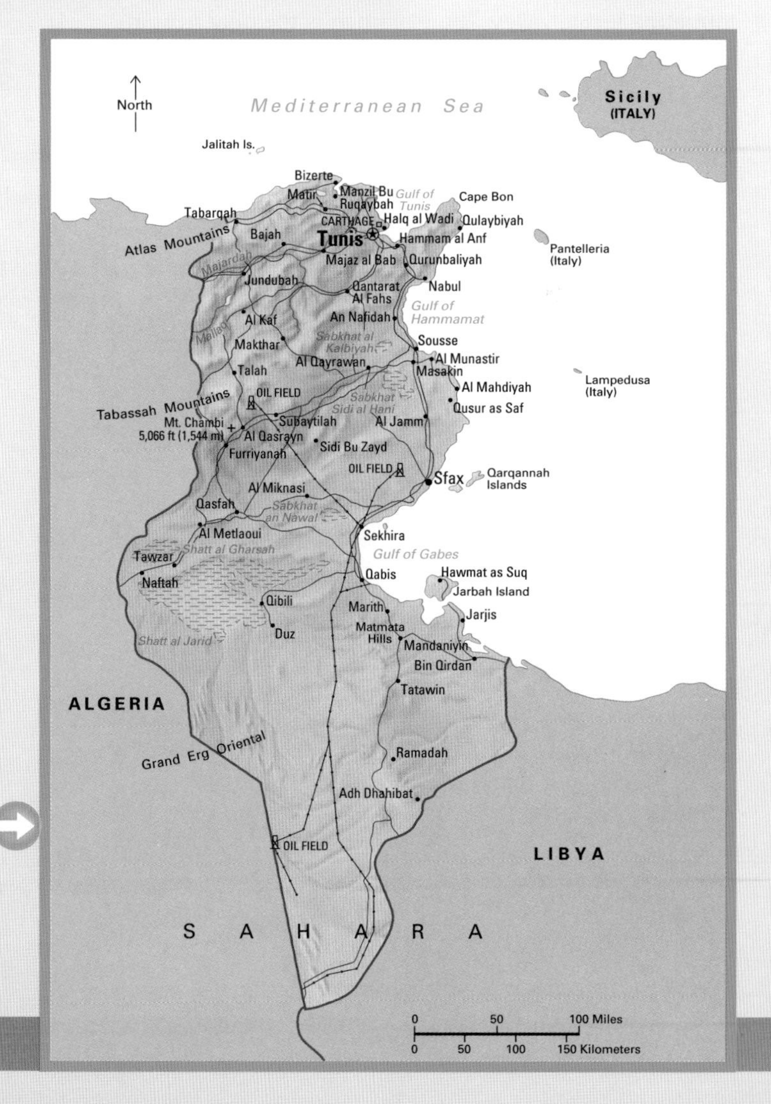

Tunisia is bordered on the north and northeast by the Mediterranean Sea, on the west by Algeria, and on the southeast by Libya. At its most northern point, it is only 85 miles (137 kilometers) from Sicily, Italy. Tunis is the nation's capital and Tunisia's largest city. It lies on the northern coast near the Gulf of Tunis. The coastal plain that extends from Tunis to Sfax includes the country's largest towns.

LAND AND ECONOMY

Tunisia is a small country with a fairly gentle landscape. It generally lacks geographic features that create divisions between people, such as high mountain regions or wide deserts. Communication between all areas and people of Tunisia is relatively easy. These conditions may be one reason that Tunisia's people lead similar lives.

Tunisia does have a northern mountain region, but the mountains are low. The region is heavily populated and prosperous. The northern mountains are two branches of the Atlas Mountain chain that extends across northwestern Africa. The northern branch is called the Atlas; the southern branch is called the Tabassah. Few of the mountain peaks reach more than 2,000 feet (610 meters). Hills and grassland lie between the ranges.

In the north, the Majardah River flows from west to east, forming a fertile valley. The Majardah is the only river in Tunisia that does not dry up in the summer. The country's most fertile farmland lies in the Majardah Valley, and much of the nation's wheat crop is grown there.

Magnificent stone statues from ancient Carthage still stand near Tunis. Carthage was founded by Phoenician traders about 800 B.C. but destroyed by Arab invaders in the A.D. 600's. The ruins of Carthage attract many tourists.

Along the northeast coast, a plain stretches from Tunis in the north to Sfax in the south. This region, too, is fertile and heavily populated. Cereals, citrus fruits, and olives are grown there. The country's largest cities and towns lie along this coast.

Northern and northeastern Tunisia has a Mediterranean climate with hot, dry summers and warm winters with moderate rainfall. Rainfall can be irregular, however, and droughts occur every three or four years. The average temperatures are 79° F (26° C) in summer and 52° F (11° C) in winter.

South of the Tabassah Mountains, the land descends to a plateau. Because the mountains block moisture-carrying winds from the northwest, the plateau receives little rainfall. Coarse grass covers the land, which is called the *steppe*. Tunisians

An oasis area near Matmatah in east-central Tunisia is marked by date palms and white stone buildings. The fertile plain of Tunisia's east coast gives way to the semidesert steppe, which ends with the arid Sahara in the south.

Young herders watch their cattle in the upland pastures of northern Tunisia. The area between the Atlas and Tabassah mountains includes the rich agricultural and livestock center along the Majardah River.

herd cattle, sheep, and goats in this region. Along the coast, farmers use the heavy dew that settles on the land to grow olives and cereals.

South of the plateau, the land descends to the Sahara, where great salt lakes called *chotts* or *shatts* lie below sea level. Valuable groves of date palms grow in the oases scattered throughout the higher ground of this arid region.

Like the fertile farmland and major cities, most of Tunisia's industries are located in the north and east. Food processing, textiles, and the manufacturing of basic consumer goods make up most of the country's industrial production.

Tunisia is a leading producer of phosphates, which are used to make fertilizers. The nation also has petroleum resources, which it uses for its own needs as well as for export. Clothing, leather products, and basic manufactured goods rank among Tunisia's most important exports. The chief agricultural products are barley, dates, grapes, olive oil, olives, and wheat.

Although Tunisia is not rich in natural resources, it has a more balanced economy than many of its neighbors in northern Africa. Tunisia's pleasant climate, beautiful beaches, historical sites, and proximity to Europe make it popular with tourists. The major attractions for visitors include Jerba Island; Saharan oases; ancient Phoenician, Carthaginian, and Roman ruins; and historic forts and mosques.

For hundreds of years, trade routes have connected Tunisia to the part of Africa that lies south of the Sahara. The French left good roads and railroads, and the Tunisian government has further improved the system since independence. The country's chief ports include Tunis, Bizerte, Sousse, and Sfax.

The Tunis metropolitan area is Tunisia's chief industrial center. Most of the country's banks and insurance companies are headquartered in Tunis, as are several international organizations.

A Tunisian worker empties a bag of olives at the start of the annual harvest in December. Olives are a major farm product in Tunisia, a crop important both for local consumption and for export.

PEOPLE

Almost all Tunisians are Arabs and followers of the Islamic religion. Because they speak the same language and follow the same religion, the people of Tunisia have a more uniform lifestyle than people in other countries of northern Africa.

The nation's history as a French colony has left its mark on Tunisian life in several areas, including architecture and food. Although Arabic is the nation's official language, many Tunisians speak French as well.

Small groups of Europeans, Jews, and Berbers also live in Tunisia, though most of the French and Italian people left after the nation gained its independence in 1956. Jewish communities had existed in Tunisia for centuries, but most Jews also left after independence.

Berbers lived in the region long before the Arabs arrived in the A.D. 600's, but most Berbers quickly adopted the Arab culture and became Muslims. Their descendants make up the majority of modern Tunisians, who claim Arab ancestry, speak Arabic, and practice Islam. Today, small Berber communities are found only on the island of Jarbah, in a few villages and oases near the edge of the Sahara, along the Libyan border, and in the northern mountains.

Entering the mosque of Al Qayrawan, a Muslim woman helps a young worshiper remove his shoes. Muslims ceremonially wash their face, hands, and feet immediately before prayer.

A covered marketplace in the city of Susah (Sousse) is crowded with customers and merchandise on a busy shopping day. Such covered marketplaces are typical in the medinas (old sections) of Tunisia's cities.

Most Tunisians live in urban areas. These people include many migrants from rural areas who came to find work and a better life in the city. The most populated cities and towns lie on the northeast coast. Tunis is the largest city as well as the nation's political and cultural center.

In most Tunisian cities, picturesque old sections contrast with new, modern areas. Narrow, crowded streets wind through the old quarters, where covered markets attract buyers and sellers. Wide, treelined avenues that were built by European colonists run through the modern sections, where European-style buildings are also common.

Many other Tunisian people live in villages or on farms. In the past, many rural people lived in mud huts and tents, but today most rural homes are made of stone or concrete. A small group of Berbers live in underground homes carved out of limestone hills in the south near the Sahara.

Many rural Tunisians still wear traditional Arab clothing—a long, loose gown or a coatlike garment with long sleeves, and a turban or skullcap. Women wear a white, full-length robe called a *hijab*. The hajib is wrapped around the body, over the head, and across the lower part of the face. Many people in the cities wear clothing similar to that worn in Europe and North America

Arab and Islamic customs form the foundation of Tunisians' traditional values. Islam governs individual actions and family life. Muslims traditionally pray five times a day, give alms to the poor, and fast. According to Arab custom, people live in extended families, with grandparents as well as parents and children. Women are largely restricted to the home, and men have great authority.

The government of Tunisia has tried to modernize the country, sometimes at the cost of these traditional beliefs and practices. Even before independence, the nuclear family, consisting of just a husband and wife and their children, had become much more common in Tunisia. Since independence, women have been given the right to vote, the right to divorce, and other powers. Some Tunisians oppose the Western influence and would like the country to return to more traditional Islamic ways.

Underground homes in the Matmata hills of southern Tunisia have sheltered Berbers for many years. Berbers carved these dwellings out of limestone hills.

Buses, pedestrians, and automobiles crowd a street in Tunis. The city is the country's cultural as well as economic and political center. Most Tunisian cities have modern sections as well as old quarters.

TURKEY

Turkey is a Middle Eastern country that lies in both Europe and Asia. About 3 percent of the country occupies the easternmost tip of Europe. This region of green, fertile hills and valleys is called Thrace. The rest of Turkey, a mountainous peninsula known as Anatolia, or Asia Minor, lies in Asia. The Bosporus Strait separates Thrace and Anatolia. This strait is a major waterway that links the Mediterranean and Black seas.

Turkey resembles the Western world in its industrial progress. In other ways, it has kept the old Asian traditions. In the industrialized cities in northern and western Turkey, factories and mills use modern techniques to process food, beverages, and textiles. But farmers on the Anatolian plateau still raise livestock as their ancestors did.

Islamic law strongly influenced Turkish life for nearly 1,000 years. However, Turkey's republican government introduced sweeping cultural and political reforms in the 1920's that discouraged or outlawed Islamic practices. Most Turkish people accepted the reforms. But many others, including those living in rural areas, resisted the changes. Diasgreement over the role of Islam in Turkish life continues to divide the nation.

Turkey has political ties with other Middle Eastern nations, but it also has strong ties to the countries of Europe. Turkey is a member of the Council of Europe and the North Atlantic Treaty Organization (NATO). Since the 1960's, Turkish workers have migrated to Europe in search of jobs. At the same time, large numbers of European tourists have been attracted to Turkey's sunny beaches on the Mediterranean Sea.

For most European visitors, the historic city of Istanbul is the gateway to the East. Situated in both Europe and Asia at the south end of the Bosporus, Istanbul is Turkey's largest city. An inlet of the Bosporus, called the Golden Horn, flows through European Istanbul. Second in size to Istanbul is Ankara, the capital city of Turkey. The city serves as a market for the grain, Angora wool and mohair, and other farm products of the area.

Nearly 70 percent of Turkey's people live in Istanbul, Ankara, and other cities and towns throughout the country. The number of urban dwellers has grown rapidly since the mid-1900's. At the same time, several minority groups have stayed in the more isolated, mountainous regions.

TURKEY TODAY

Modern Turkey came into being when the country was proclaimed a republic in 1923. Under the new government, Turkey began to change from an almost entirely agricultural country to a more industrialized nation. Today, the value of its industrial production is greater than that of its agricultural output. Yet, industrialization is only one development that makes Turkey what it is today.

A changing way of life

Turkish society began to change dramatically when Turkey became a republic on Oct. 29, 1923. Mustafa Kemal became the country's first president. Kemal, renamed Atatürk (meaning father of the Turks), was a hero in the struggle for national independence against the Greek and Allied forces. He served as president until his death in 1938.

Atatürk launched a national program to sweep away centuries-old customs and introduce modern ways. He declared that Islam would not be the state religion, and he discouraged or forbade many rules of Islamic tradition. He gave women more civil rights and abolished such customs as the wearing of the veil for women and the fez for men.

Atatürk introduced political reforms too. He abolished the offices of the sultan and the caliph, both strongholds of Islamic political leadership. Atatürk and his government also developed the modern Turkish language, which was easier for people to learn.

These social, religious, and political reforms helped make Turkey a modern nation. However, such drastic changes in the Turkish way of life caused much tension. Many Turks would like to return to Islamic traditions.

Recent political developments

After many years of instability in Turkey, Army leaders took control of the government in 1980. Turkey adopted a new constitution in 1982 and returned to civilian rule in 1983, when parliamentary elections were held.

In 1993, the National Assembly elected Süleyman Demirel as president, and Tansu Çiller became Turkey's first woman prime minister. In 1996, Necmettin Erbakan became prime minister. He was the first person from an Islamic party to head the government since Turkey became a republic in 1923. But in 1997, Erbakan resigned under pressure from

FACTS

Official name:	Turkiye Cumhuriyeti (Republic of Turkey)
Capital:	Ankara
Terrain:	Mostly mountains; narrow coastal plain; high central plateau (Anatolia)
Area:	302,535 mi² (783,562 km²)
Climate:	Temperate; hot, dry summers with mild, wet winters; harsher in interior
Main rivers:	Euphrates, Kizil, Sakarya, Büyükmenderes, Aras
Highest elevation:	Mount Ararat, 16,946 ft (5,165 m)
Lowest elevation:	Mediterranean Sea, sea level
Form of government:	Republic
Head of state:	President
Head of government:	Prime minister
Administrative areas:	81 iller (provinces)
Legislature:	Turkiye Buyuk Millet Meclisi (Grand National Assembly of Turkey) with 550 members serving five-year terms
Court system:	Constitutional Court, Court of Appeals
Armed forces:	510,600 troops
National holiday:	Republic Day - October 29 (1923)
Estimated 2010 population:	76,606,000
Population density:	253 persons per mi² (98 per km²)
Population distribution:	68% urban, 32% rural
Life expectancy in years:	Male, 70; female, 74
Doctors per 1,000 people:	1.6
Birth rate per 1,000:	19
Death rate per 1,000:	6
Infant mortality:	23 deaths per 1,000 live births
Age structure:	0-14: 27%; 15-64: 67%; 65 and over: 6%
Internet users per 100 people:	3
Internet code:	.tr
Languages spoken:	Turkish (official), Kurdish, other minority languages
Religions:	Muslim 99.8% (mostly Sunni), other 0.2%
Currency:	New Turkish lira
Gross domestic product (GDP) in 2008:	$794.23 billion U.S.
Real annual growth rate (2008):	1.5%
GDP per capita (2008):	$10,615 U.S.
Goods exported:	Chemicals, iron and steel, machinery, motor vehicles, petroleum
Goods imported:	Chemicals, food, iron and steel, machinery, transportation equipment
Trading partners:	France, Germany, Italy, Russia, United States

the military. Mesut Yilmaz replaced him as prime minister. Following elections held in 1999, former Prime Minister Bülent Ecevit became head of a coalition government.

In 2002, Abdullah Gül became prime minister. Recep Tayyip Erdogan replaced him in 2003 and remained prime minister after his party won parliamentary elections in 2011. Gül became president in 2007. Also in 2007, Turkish voters approved constitutional changes that included direct presidential elections.

Population trends

Turkey was almost entirely an agricultural country when it became a republic in 1923. Since that time, the population of Turkey has quadrupled. Today, more than two-thirds of the people live in urban areas. The migration to the cities has caused overpopulation and overcrowding.

During the 1920's and 1930's, the Turkish government tried to bring the Kurds and other tribal people into the mainstream of modern Turkish life. These people, who had for many years lived in tribal groups as nomads or in isolated communities, initially resisted. Since then, some Kurds have adopted modern Turkish culture. In 1995, however, Turkey invaded northern Iraq and attacked Kurdish camps, which it accused of harboring guerrillas who committed acts of violence within Turkey.

In August 1999, a powerful earthquake struck northwestern Turkey. The quake killed more than 17,000 people.

The tomb of Kemal Atatürk stands in Ankara, the capital of Turkey. Atatürk was the founder of the Republic of Turkey. He served as its first president.

Turkey lies in both Europe and Asia. Part of its northwestern region lies in Europe. Yet much of Turkey's history and its Islamic tradition place it firmly in Asia. Its strategic location between the Mediterranean and Black seas has played an important role in its history. Turkey has been a center of trade for centuries.

ENVIRONMENT

Turkey is a land of broad, fertile river valleys and arid highlands. Turkey also has several saltwater lakes and numerous rivers. Its rugged terrain of high mountains and broad plateaus often makes travel and communication difficult. Despite these challenges, the people of Turkey have used the land well, as the population has grown to more than 75 million people.

Landscape

Turkey can be divided into eight land regions: the Northern Plains, the Western Valleys, the Southern Plains, the Western Plateau, the Eastern Plateau, the Northern Mountains, the Southern Mountains, and the Mesopotamian Lowlands.

The Northern Plains cover Thrace and extend along the Black Sea coast of Anatolia. Their gently rolling grasslands are ideal for farming and grazing. This region yields crops of corn, fruits, nuts, tea, and tobacco.

The conelike formations at Goreme were caused by centuries of wind and water wearing away at volcanic rock. Today, the formations are a major tourist attraction.

The Western Valleys are another important region for agriculture. These broad, fertile valleys lie along the coast of the Aegean Sea. The rich soil produces barley, cotton, olives, tobacco, and wheat.

The Southern Plains also produce a great variety of crops. Farmers in this region grow cereal grains, citrus fruits, and cotton.

The Western Plateau is a region of highlands and scattered river valleys. It extends across central Anatolia and receives very little rainfall. As a result, most of the land is not cultivated. It is used mainly for livestock grazing.

The Northern, or Pontic, Mountains rise between the Northern Plains and the Anatolian plateau. The Southern Mountains consist of the Taurus Mountains and several smaller ranges on the southern edge of the Anatolian plateau.

The island of Aktamar is known for its Armenian Church of the Holy Cross. Aktamar is situated near the shore of Lake Van in eastern Turkey.

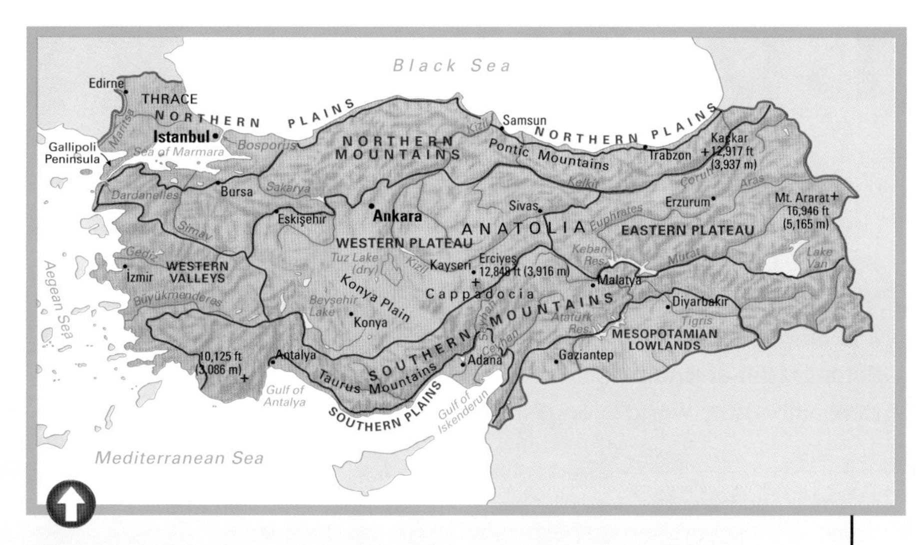

The Anatolian peninsula is a high plateau surrounded by mountains. The clouds and rainfall that form over the mountains evaporate over the arid steppes of the Anatolian interior. Only the coastal areas enjoy a warm Mediterranean climate.

Volcanoes and earthquakes are common along the fault lines of the Northern and Southern Mountain regions. The cone-shaped rock formations near Goreme are the remains of volcanic deposits that have been eroded by wind and water. Turkey's largest body of water, Lake Van, was formed when a large volcano prevented water from draining off what is now the lake bed.

The historic Tigris and Euphrates rivers have their source in the Eastern Plateau region. The rivers flow down into the Mesopotamian Lowlands. The fertile plains and river valleys of the Mesopotamian Lowlands are ideal for producing abundant crops of cereal grains and fruits.

Climate

Because of its location on the Mediterranean Sea, many people believe that Turkey would have a sunny, mild climate. However, the Mediterranean climate, with its warm summers and mild winters, can only be found in Thrace and along the southern and western coasts of Anatolia. The rest of Anatolia is far too high in altitude for such mild conditions.

Northeastern Turkey has mild summers but bitterly cold winters. Southeastern Turkey and the interior of Anatolia have cold winters with heavy snowstorms.

Yearly rainfall in coastal areas averages from 20 to 30 inches (51 to 76 centimeters) along the Aegean and Mediterranean to more than 100 inches (254 centimeters) near the Black Sea.

Mount Ararat, an extinct volcanic massif, is Turkey's highest mountain. A massif is formed when a large block of Earth's crust shifts up or down.

ISTANBUL

Istanbul is the largest city in Turkey. It is the only major city located on two continents—Europe and Asia. The Bosporus, a strait in northwestern Turkey that connects the Black Sea and the Sea of Marmara, separates Asian and European Istanbul. The Bosporus Bridge and the Fatih Sultan Mehmet Bridge link the two sections. Both bridges rank among the world's longest suspension bridges.

Ancient sights

Some of Istanbul's most spectacular sights can be seen from the Bosporus Bridge. Beautiful *mosques* (Islamic houses of worship) tower skyward and form a dramatic skyline. These large, elaborate structures are noted for their towers, called *minarets.*

Hagia Sophia is one of Istanbul's most famous landmarks. Built between A.D. 532 and 537, it is the world's most impressive surviving example of Byzantine architecture. The building is known for its huge central dome and its richly decorated interior. Hagia Sophia served as a cathedral and then a mosque. It became a museum in 1935.

The tea gardens of Istanbul are popular places where Turks meet and discuss the business of the day. Here, a lottery-ticket seller works his trade.

The city's landmarks reflect its ancient origins. At various times, Istanbul, also known by the earlier name Constantinople, served as the capital of the Roman, Byzantine, and Ottoman empires. Today, the glories of these empires can still be seen in Istanbul.

Hagia Sophia may be the most stunning example of Byzantine architecture in Istanbul. It was built as a Christian cathedral by the Emperor Justinian I in the A.D. 530's. Its name comes from the Greek phrase meaning *holy wisdom*. Sometimes Hagia Sophia is called Saint Sophia. The marble-lined

walls of the cathedral have many colors and designs, and mosaics decorate some of the walls and floors. It was converted into a mosque in 1453. Since 1935, Hagia Sophia has served as a museum.

Traces of the Eastern Roman Empire survive in soaring *aqueducts* (channels built to carry water from one place to another) and underground *cisterns* (reservoirs). The largest of these cisterns, the *Yerebatan Sarayi* (Sunken Palace), has been restored so that visitors can walk through its huge vaulted chamber.

When a group called the Ottomans conquered Constantinople in 1453, they established the Ottoman Empire. Istanbul was its capital. The Ottomans built the Topkapi Palace, a group of elaborate buildings, courtyards, and gardens. Begun in 1462, it became the residence of the sultans until the founding of the republic in 1923. Topkapi Palace is now a museum that showcases Ottoman treasures. There, visitors can see the Chamber of the Sacred Relics, which contains strands of the Prophet Muhammad's beard.

Everyday life

Today, Istanbul is a modern, bustling city. Taxis called *dolmus* scurry through traffic. Shoppers and merchants pack the Grand Bazaar, a maze of narrow arcades containing thousands of shops. Often, people stop in the *hamams,* or Turkish baths, to freshen up after a busy day.

The waters of Istanbul are as busy as its streets. The Bosporus and the Sea of Marmara are always crowded with ferries, water-buses, and ships.

The busy Bosporus Strait divides Istanbul's European and Asian sections. Ferries cross the strait each day. Süleymaniye Mosque towers in the background.

PEOPLE AND AGRICULTURE

The coastal valleys of Turkey, with their rich soil and mild climate, are the most productive farmlands in the country. Here, Turkish farmers grow fruits, grains, nuts, cotton, and tobacco. They also raise sheep, goats, and other livestock. Wheat and barley are grown in the desertlike conditions of the Anatolian plateau.

Much of the land in Turkey is unsuitable for farming. Since the 1800's, the need to feed a growing population has forced the inhabitants of Anatolia to clear forests for more farmland. But in recent years, the Anatolians have realized the value of their forests. They have now begun fencing in the forests and planting new ones.

Farming

Many Turkish farmers still use traditional farming methods. Tractors and other machines would help them work the land better, but most Turkish farms are too small for such equipment. Instead, teams of oxen, hook plows, threshing machines, and hand sickles are common sights.

Turkish farmers grow cotton for both fiber and for cottonseed oil. Agriculture accounts for only about 15 percent of the value of all goods and services produced in Turkey, but agriculture employs about 45 percent of the country's people.

Few crops can grow in the dry Anatolian plains without extensive irrigation to provide water. Many of the region's people, like these herdsmen, raise livestock.

A small workshop in the town of Alanya allows this silk weaver to escape rural poverty. Many villagers still keep up their craft skills at home.

Turkish farmers, including women, work long hours in the field. Even so, their yield per acre is well below that of Western Europe or North America. Most Turkish farmers just want to feed themselves and their families. They sell any surplus crops in the local market. Sometimes the farmers add to their incomes by raising chickens or other small animals.

Village life

Turkish farmers live in small villages in the countryside. Though modern conveniences such as electricity and paved roads have come to the villages, life remains simple. Many reforms of the 1900's were designed to modernize Turkish life, but the old ways have changed little for Turkey's villagers.

Everyday life in the Turkish villages is centered around Islamic rules and traditional family ties. Each village has its own mosque, as well as a bust or statue of Atatürk. Daily meals consist of simple, hearty dishes of lamb, bread, and vegetables. Clothing in many villages still follows the Islamic tradition, even though the government has discouraged it. The men's loose-fitting baggy trousers and women's headscarves are reminders of the old ways.

Crafts play an important part in village life. For hundreds of years, Turkish craft workers have made colorfully decorated ceramic dishes, bowls, and other objects. Their handcrafted ceramic tiles have decorated mosques and palaces throughout the country.

Little by little, some of the old customs have been mixed with the new. The tea house, a traditional gathering place in the village, now often has a television and city newspapers. The country villages are no longer as isolated as they once were.

Migration to cities

Many rural people have been forced to leave their homes to seek jobs in the more industrialized areas. People in the mountainous regions are often away from their villages for long periods. Thousands of them leave their homes for Cukorova in southern Turkey, where the farming brings them seasonal work.

Some workers have also left the rural areas for jobs in the large industrial cities of Istanbul, Izmir, Bursa, and Adana. Others have even left the country. About a million Turks now live abroad as gastarbeiter (guest workers), mostly in Europe. In Germany, more than half a million Turks are guest workers.

ECONOMY

Turkey is a developing country. When it became a republic in 1923, Turkey's economy was almost entirely based on agriculture. At that time, Turkey had only 118 factories. Under the government's direction, the number of factories increased to more than 1,000 by 1941. Turkey now has several thousand factories. Today, the value of the country's industrial production is about twice that of its agricultural output.

Although Turkey has become a more industrialized nation, progress has been uneven. The western regions of the country have seen a growth in the job rate and in private investments. The eastern regions, though, have been slower to adopt new technology.

Water pours from the Atatürk Dam. The dam stands on the Euphrates River on the border of Adıyaman Province and Sanlıurfa Province in southeastern Anatolia. The dam was completed in 1990 to generate electricity and provide irrigation.

The government has long been heavily involved in many aspects of Turkey's economy. The government has owned much of the country's transportation and communications industries, and it has controlled other industries as well. However, private companies have become increasingly important. In the 1980's, the government began a program to reduce its control of industries and to allow more private ownership.

Areas of industry

Turkey's chief industrial area is the Marmara region, which includes the cities of Istanbul, Izmir, Adapazari, and Bursa. More than half of Turkey's manufacturing centers are located there. This region shows rates of population growth far above the national average.

Istanbul is Turkey's most important center for commerce and industry. For centuries, it has served as a major seaport. The Bosporus Bridge, which connects the Asian and European sections of Istanbul, is now an important link in the trade route between Europe and the Middle East.

Bursa, once the capital of the Ottoman Empire, has become a major center for car and truck factories. Other

The container port at Haydarpasa near Istanbul on the Bosporus, handles cargo bound for the industrial towns around the Sea of Marmara.

Millions of tourists visit Turkey every year. These tourists are near the Temple of Hadrian in Ephesus, on Turkey's west coast.

A textile mill, represents one of Turkey's biggest industries. Cotton and wool fabrics are major exports.

commercial centers include Greater Ankara, the Cukorova area around Adana, the Aegean port city of Izmir, and the Black Sea ports of Zonguldak and Samsun.

Together, these areas are home to a large percentage of Turkey's urban population. The areas also draw a majority of the migrant workers. Although the migrant workers may lose some of their traditional family ties, they enjoy a higher standard of living and better educational opportunities by living in the city.

Agriculture and food processing

Farming provides jobs for just under half of Turkey's workers. In most years, Turkey's farmers produce enough food for all the people, plus a surplus to sell abroad. Many of the farms in Turkey are small and use traditional farming methods. The rest are larger farms and cotton fields, which attract migrant laborers.

One of Turkey's largest manufacturing industries is food processing. The industry has proven to be an excellent way for Turkey to link its two major areas of production—agriculture and manufacturing. By processing the food as well as growing it, Turkey has created more ways to increase its economic output. Food processing has also created many jobs for rural migrants as well as for seasonal workers.

One of Turkey's biggest areas of economic growth is tourism. Visitors from all over the world come to enjoy the sunny beaches and sea breezes of the Turkish Riviera along the Mediterranean and Aegean coasts.

HISTORY

The history of Turkey dates back thousands of years. Archaeologists have found evidence of an advanced society inhabiting what is now Turkey before 6000 B.C. Turkey's position between the Mediterranean and Black seas has made it a crossroads for migrating people and conquering armies throughout history.

Empires and conquerors

The first people of Turkey to be recorded by history were the Hittites. By 1500 B.C., the Hittites had a powerful empire and were the leading rulers of the Middle East.

Three hundred years later, large areas of Anatolia fell to the Phrygians, the Lydians, and other peoples. Meanwhile, the Greeks founded the city-states of Ephesus, Pergamum, and others along the Aegean coast.

Between about 550 and 513 B.C., the Persian Empire took control of Anatolia and Thrace. They ruled the region until Alexander the Great of Macedonia conquered them in 331 B.C. After Alexander died in 323 B.C., the area became a battleground, with his successors fighting for power among themselves.

Anatolia did not find peace until 63 B.C., when the Roman general Pompey conquered the region and declared it part of the Roman Empire. In A.D. 330, Emperor Constantine I, also known as Constantine the Great, moved the capital of the Roman Empire to Byzantium in Thrace. Byzantium was renamed Constantinople. The Byzantine Empire thrived for several hundred years as a center of ancient Greek literature and philosophy.

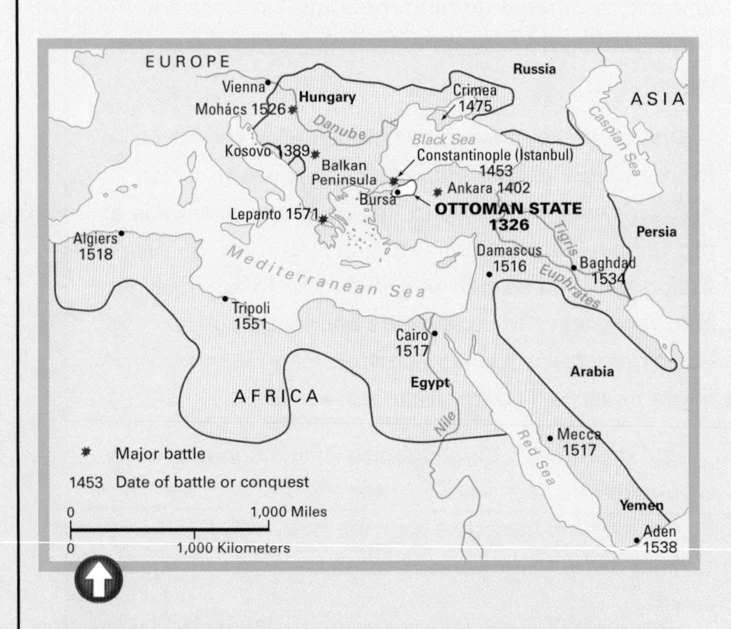

The Ottoman Empire began during the 1300's as a small state around the city of Bursa. It grew to include much of the Middle East and parts of northern Africa and southeastern Europe.

Dressed in the colorful costumes of the Ottoman Empire, these musicians re-create the elaborate entertainment enjoyed by the sultan and his court.

In 1071, the Muslim Seljuk Turks defeated the Byzantine army and set up their own empire. The Seljuk Empire survived until 1243, fighting off the Christian crusaders who wanted to drive the Turks from the region.

The rise of the Ottomans

In 1243, the Seljuk Empire was destroyed by invading Mongols. But the Mongol Empire soon fell apart. In 1326, a Turkish tribe known as the Ottomans seized the city of Bursa. In 1453, Ottoman forces led by Mehmet II captured Constantinople and renamed it Istanbul.

The Ottoman Empire reached its peak under the reign of Sultan Suleyman I, called "the Magnificent," who ruled from 1520 to 1566. Under his leadership, the Ottomans extended their empire to Yemen in the south, Morocco in the west, and Persia in the east.

The Ottoman Empire began to decline in 1571. By the 1800's, the empire came to be known as "the sick man of Europe." In 1821, Greek nationalists revolted against Ottoman rule and won their independence.

In 1908, a group of Turkish students and military officers known as the Young Turks led a revolt against

TIMELINE

2000 B.C.	Hittites migrate to Anatolia.
1200 B.C.	Greeks found city-states on Aegean.
63 B.C.	Roman General Pompey conquers Anatolia.
A.D. 330	Constantine the Great moves capital of Roman Empire to Byzantium.
1000's	Seljuk Turks invade Anatolia.
1071	Seljuk Turks defeat Byzantines at Manzikert.
1096-1099	Christian Crusaders defeat Seljuk Turks in western Anatolia.
1243	Asian Mongols invade Seljuk Empire.
1300's	Ottomans begin to build an empire.
1326	Ottomans seize Bursa.
1453	Ottomans capture Constantinople, ending Byzantine Empire.
1520-1566	Reign of Suleyman I, *the Magnificent,* height of Ottoman Empire.
1529	European forces defend Vienna against Ottoman attack.
1571	Europeans defeat Turkish fleet at Lepanto.
1783-1914	Ottoman Empire loses territory in series of wars.
1821-1829	Greeks rebel against Ottoman rule.
1908	Revolt of the Young Turks.
1914-1918	Turkey sides with Germany in World War I.
1915	Turks defend Dardanelles against Allied attack.
1920	Treaty of Sèvres divides Ottoman Empire.
1923	Turkey becomes a republic under President Kemal Atatürk, who begins social-reform movement.
1925	Kurdish revolt fails.
1947	United States gives military aid to Turkey.
1960	Turkish army overthrows government.
1961	Free elections held.
1964	Hostilities between Turks and Greeks in Cyprus.
1980	Turkish army takes control of government.
1982	New constitution adopted, and General Kenan Evren named president.
1993	Tansu Çiller becomes Turkey's first woman prime minister.
1999	Earthquake kills more than 17,000 people in northwestern Turkey
2003	Recep Tayyip Erdogan becomes prime minister.
2007	Abdullah Gul becomes president.

Sultan Mehmet II (1432-1481)

Sultan Suleyman I, *the Magnificent* (1494-1566)

Kemal Atatürk (1881-1938)

the sultan. They demanded that the Ottoman Empire be restored to its former greatness. But by then, many Turkish people no longer cared about keeping an empire.

The final blow to the Ottomans came when they entered World War I (1914–1918) on the side of Germany and Austria-Hungary. The Ottoman army fought long and hard against British and French troops to keep control of the Dardanelles. This narrow strait, which links the Aegean Sea with the Sea of Marmara and the Bosporus, was an important strategic waterway. Although the Ottomans prevented the Allies from controlling the Dardanelles, they lost the war to them in 1918.

The new Turkey

After World War I, Greek troops with Allied support tried to take over Anatolia. Mustafa Kemal rallied the Turks and drove out the invaders. Kemal became president of the new Republic of Turkey in 1923. He was given the new surname of Atatürk, which means *father of the Turks.*

As president, Atatürk introduced many social reforms because he believed that Turkey had to change its old ways to survive. Atatürk's hopes for a strong, unified nation grew into the foundation of modern Turkey.

Whirling dervishes are members of certain Islamic religious orders. They devote themselves to prayer and other forms of devotion. Members dance and whirl as part of their worship.

HISTORIC SITES

The archaeological sites and ancient buildings in Turkey stand as reminders of the land's long and eventful history. From the elaborate mosques of Istanbul to the scattered ruins of Mount Nemrut, each relic is a part of the rich cultural heritage of Turkey.

Istanbul

When the Ottoman Sultan Mehmet II captured Constantinople in 1453, he refused to plunder the city's artistic treasures. Instead, he converted the jewel of Byzantine architecture, Hagia Sophia, from a church into a mosque. Today, Hagia Sophia is a reminder of the different religious groups that have influenced Turkish history.

The Süleymaniye Mosque and the Sultan Ahmet Mosque are two masterpieces of Turkish architecture. The Süleymaniye Mosque was completed in 1557. The Sultan Ahmet Mosque—better known as the Blue Mosque, for the blue-glazed tiles that cover its inside walls—was begun in 1609 and took seven years to complete.

Like the Süleymaniye Mosque and the Blue Mosque, the Topkapi Palace was built during the Ottoman Empire. In 1840, the Sultan Abdul Mecit commissioned a new palace, the Dolmabahçe Saray, on the Bosporus.

Bursa

Across the Sea of Marmara and south of Istanbul lies the ancient city of Bursa. The Ottomans made Bursa their first capital when they defeated the Byzantines in 1326. The city has kept its Ottoman flavor to this day. Bursa's most important landmark is the Ulu Cami (Great Mosque).

South of Istanbul and Bursa on the Aegean shore lie the ruins of the ancient city of Pergamum. Pergamum was an important center of Greek culture during the 200's B.C. South of Pergamum lies Ephesus, a major trading and banking center under the Greeks in the 300's B.C. Ephesus is most famous for its Temple of Artemis, dedicated to the Greek goddess Artemis. The Temple of Artemis is one of the Seven Wonders of the Ancient World.

The Arcadiane runs through the heart of the Greek city of Ephesus, between its former harbor and the stone theater. Ephesus became a major trading center in the 300's B.C.

Eastern Turkey

High on the Anatolian plateau lies Erzurum, once an important point on the Silk Road from China. East of Erzurum stands Kars. Set on a rocky peak, its citadel was once an Armenian palace.

On the southeastern shore of the Black Sea, near the ancient town of Trabzon, stands the Greek Orthodox Monastery of Sumela, founded in the 500's. The monastery is built into the walls of a cliff 4,100 feet (1,250 meters) above sea level. Now abandoned and crumbling away, it is the most important Byzantine monument in the region.

To the south of Trabzon lies Mount Nemrut, where the Greek ruler Antiochus I (62-32 B.C.) built huge statues of Greek and Persian gods. Remains of these statues lie among the rocks, scattered by earthquakes, wind erosion, and storms.

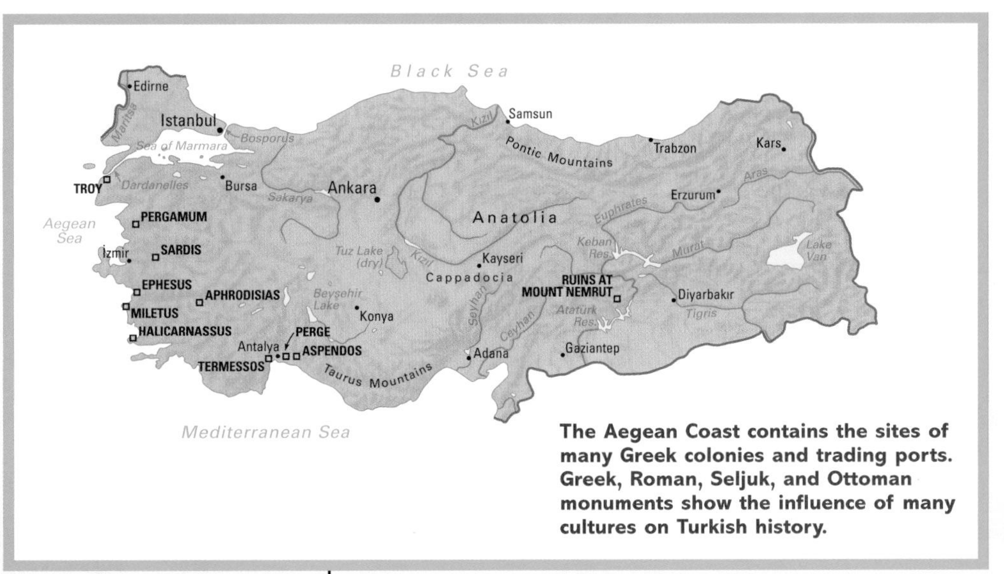

The Aegean Coast contains the sites of many Greek colonies and trading ports. Greek, Roman, Seljuk, and Ottoman monuments show the influence of many cultures on Turkish history.

Central Turkey

The ancient city of Konya, in central Turkey, contains some of the world's finest examples of Seljuk art and architecture. Konya is also home to the famous Muslim Sufi (mystic) Rumi (1207-1273). Rumi was the founder of the whirling dervishes, an order of Muslim mystics who whirl and dance to the music of a reed pipe as part of their worship.

Hagia Sophia was built by the Byzantine Emperor Justinian I. Over the center is a great dome 185 feet (56 meters) above the floor.

THE TURKISH RIVIERA

The natural beauty and sunny climate of Turkey's southern coast have made the area a popular destination for vacationers. This beautiful stretch of shoreline—between Bodrum on the Aegean Sea and Antioch (also known as Antakya) near the Syrian border—has earned the name *Turkish Riviera*. Major resort areas include Marmaris on the Aegean Coast and Antalya on the Turquoise Coast.

The Turkish Riviera attracts millions of foreign visitors every year. Tourists and sailors are drawn to the beautiful sandy beaches and deep blue waters of the Mediterranean Sea. They also come to see the archaeological sites and ancient monuments that tell the history of the region.

Although many hotels now dot the skyline of the southern coast, the region is still much as it was in ancient times.

Resort areas

For a holiday of sun, sea, and sand, many vacationers choose the fashionable resort area of Marmaris. Visitors often rent boats to explore some of the secluded beaches around Marmaris.

On the Turquoise Coast, the beaches along the Gulf of Fethiye also attract tourists. The sandy lagoon of Ölü Deniz, eight miles (14 kilometers) from Fethiye, offers some of the area's finest beaches.

Fethiye, on the Turquoise Coast, features some of the Turkish Riviera's finest beaches. The bright blue waters of the Mediterranean Sea give the Turquoise Coast its name.

Near the ruins of an ancient theater in Bodrum stands the Castle of St. Peter, dating from 1402. Bodrum also is the site of the Mausoleum of Halicarnassus.

The Turkish Riviera stretches from Bodrum on the Aegean coast to Antakya (Antioch), near the Syrian border. With its mild climate and sandy beaches, the Turkish Riviera attracts many tourists.

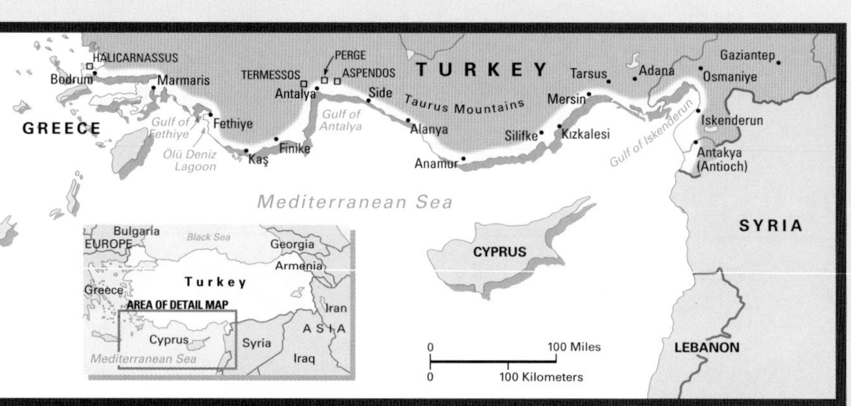

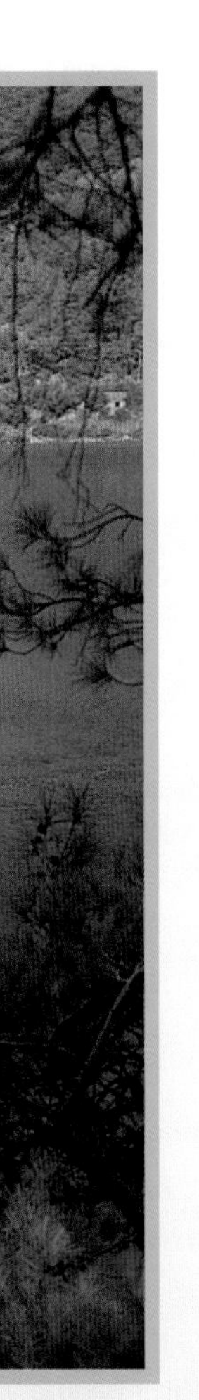

Farther south is the major tourist center of Antalya. A thousand years ago, Antalya was the winter capital of Seljuk sultans. Today, Antalya welcomes visitors to its fine beaches, charming harbor town, and bazaars.

Antalya stands on a bay enclosed by the Taurus Mountains. It enjoys daytime temperatures of more than 70° F (20° C) in March and April. Even in winter, temperatures along the coast average a mild 50° F (10° C). Diners enjoy *izgara* (charcoal-grilled) fish, a local specialty in Antalya's famous restaurants.

Glories of the past

Along with its beauty and charm, Antalya is the site of the Gate of Hadrian, a decorative gate that was built in honor of the Roman Emperor Hadrian's visit to the city in A.D. 130. It is one of the many historic sites along the Turkish Riviera.

Northeast of Antalya, the ancient city of Termessos is nestled on a mountainside at an elevation of some 5,400 feet (1,650 meters). Termessos was founded by settlers from Pisidia in the 7th century B.C. By the 100's B.C., many Greeks had settled there.

The Termessians built such a stronghold to protect their city that it is said even Alexander the Great failed to conquer it. The ruins include an *odeon* (theater) and an agora (marketplace). Along the slopes to the east, west, and south are large numbers of rock tombs.

At nearby Perge, visitors pass through an elaborate main gate built during Roman times. The visitors then enter a vast courtyard filled with statues and monuments.

East of Perge is Aspendos, the site of the best-preserved Roman theater in the Mediterranean. Even the stage has survived from ancient times, with only some restoration of the theater done by the Seljuks in the A.D. 1200's. Plays are still staged in this theater during a September festival.

East of the Gulf of Antalya lies Anamur, home to *Anamur Kalesi* (Anamur Castle). This castle was built by a Seljuk chief around 1230. Anamur Castle is one of many castles that stand along the rocky slopes of the Taurus foothills.

The Mausoleum of Halicarnassus is the Turkish Riviera's most famous historic site and one of the Seven Wonders of the Ancient World. The Mausoleum is located in Bodrum, site of the ancient city of Halicarnassus. The city's most glorious period came during the reign of King Mausolus of Caria, from 377 to 353 B.C. When King Mausolus died, his wife and sister built the Mausoleum for his tomb.

The people of the Turkish Riviera still lead a quiet life despite the growing presence of tourists. The mild climate and fertile soil are ideal for farming.

TURKMENISTAN

Turkmenistan is a country in west-central Asia. About three-fourths of it is covered by the Karakum, one of the largest deserts in central Asia. The Karakum—formed by sandy deposits from a river called the Amu Darya—consists mostly of flat clay plains, salt basins, and sand mounds. Today, the Amu Darya and the Karakum Canal provide Turkmenistan with most of its water. The Karakum Canal, which spans 850 miles (1,350 kilometers), was built during the 1950's and 1960's.

Most of Turkmenistan's people live along the Amu Darya, Murgab, and Tedzhen rivers and in the oases at the foot of the Kopet-Dag Mountains. About 85 percent of the people are ethnic Turkmen—Turkic-speaking Sunni Muslims. Some Turkmen are nomads who roam the land with their livestock and follow age-old tribal customs. The remainder of the population consists of Armenians, Kazakhs, Russians, Tatars, Ukrainians, and Uzbeks.

Many men and women wear the national dress. The women's garments—bright silk or satin dresses with elaborate embroidery and accented by elegant brooches, beads, and earrings—are especially striking.

Some Turkic tribes had settled in the Turkmenistan area by the A.D. 900's. The Mongols under Genghis Khan conquered the region in the 1200's. Between the 1400's and the 1600's, the southern part of Turkmenistan was ruled by the Persian Safavids. By 1885, Russia had control of all Turkmen lands.

Following the October Revolution in Russia in 1917, two opposing forces struggled for control of the Turkmenistan region. Local rebel forces fought against the Communist Red Army. The rebels received aid from the United Kingdom because the British feared the Red Army's southward advance toward India, then a British crown colony. By 1920, however, the Red Army had taken over the region. In 1924, the Turkmen Soviet Socialist Republic was formed.

Turkmenistan remained under the strict control of the Soviet Union's central government for many years. Finally, in the late 1980's, a movement for independence began to gain strength. In the midst of political

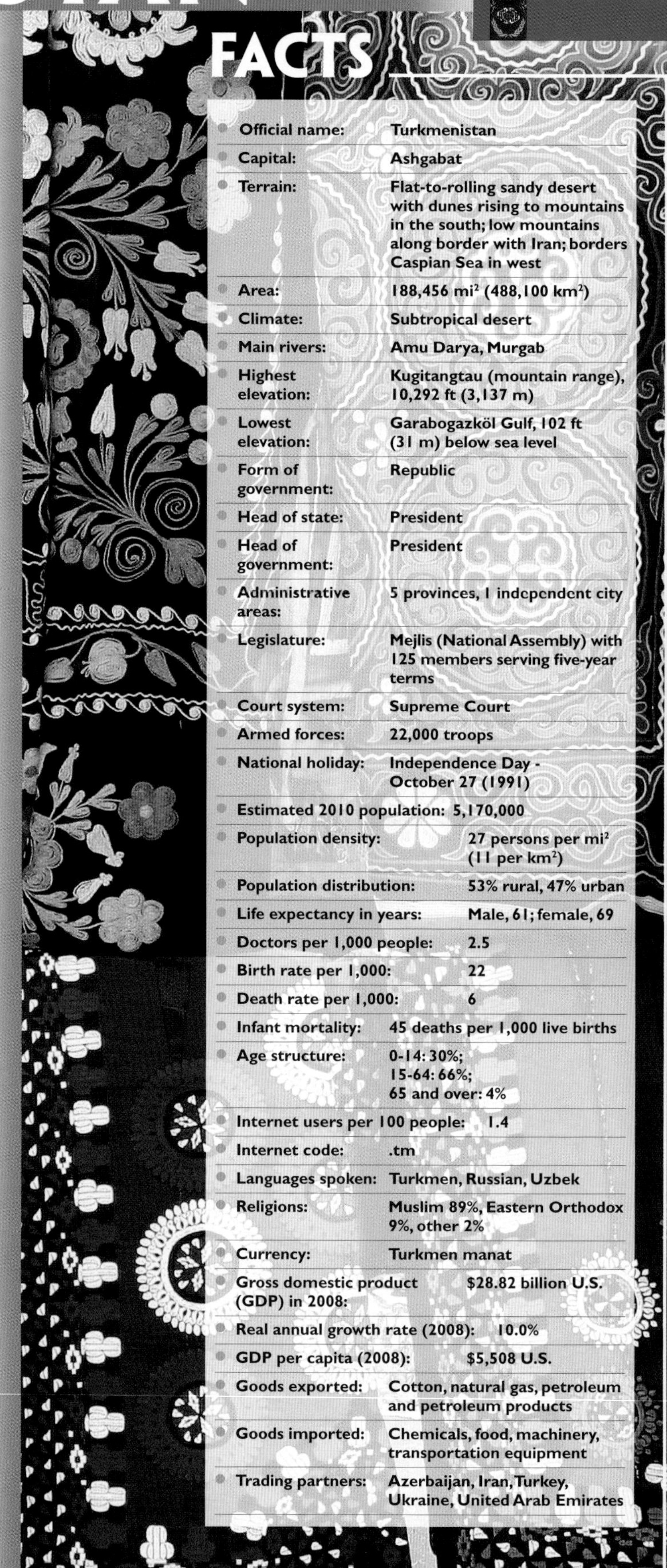

FACTS

Official name:	Turkmenistan
Capital:	Ashgabat
Terrain:	Flat-to-rolling sandy desert with dunes rising to mountains in the south; low mountains along border with Iran; borders Caspian Sea in west
Area:	188,456 mi² (488,100 km²)
Climate:	Subtropical desert
Main rivers:	Amu Darya, Murgab
Highest elevation:	Kugitangtau (mountain range), 10,292 ft (3,137 m)
Lowest elevation:	Garabogazköl Gulf, 102 ft (31 m) below sea level
Form of government:	Republic
Head of state:	President
Head of government:	President
Administrative areas:	5 provinces, 1 independent city
Legislature:	Mejlis (National Assembly) with 125 members serving five-year terms
Court system:	Supreme Court
Armed forces:	22,000 troops
National holiday:	Independence Day - October 27 (1991)
Estimated 2010 population:	5,170,000
Population density:	27 persons per mi² (11 per km²)
Population distribution:	53% rural, 47% urban
Life expectancy in years:	Male, 61; female, 69
Doctors per 1,000 people:	2.5
Birth rate per 1,000:	22
Death rate per 1,000:	6
Infant mortality:	45 deaths per 1,000 live births
Age structure:	0-14: 30%; 15-64: 66%; 65 and over: 4%
Internet users per 100 people:	1.4
Internet code:	.tm
Languages spoken:	Turkmen, Russian, Uzbek
Religions:	Muslim 89%, Eastern Orthodox 9%, other 2%
Currency:	Turkmen manat
Gross domestic product (GDP) in 2008:	$28.82 billion U.S.
Real annual growth rate (2008):	10.0%
GDP per capita (2008):	$5,508 U.S.
Goods exported:	Cotton, natural gas, petroleum and petroleum products
Goods imported:	Chemicals, food, machinery, transportation equipment
Trading partners:	Azerbaijan, Iran, Turkey, Ukraine, United Arab Emirates

upheaval in the Soviet Union in August 1991, Turkmenistan declared its independence.

In December 1991, Soviet President Mikhail Gorbachev and Russian President Boris Yeltsin agreed to dissolve the Soviet Union. The Commonwealth of Independent States (CIS) had been established the week before as a loose confederation of former Soviet republics. Turkmenistan agreed to join the CIS.

Saparmurad A. Niyazov was elected the first president of independent Turkmenistan in 1992. Gurbanguly Berdimuhammedov became president after Niyazov's death in 2006. In 2008, Turkmenistan adopted a new constitution that increased the power of the legislature. Berdimuhammedov was reelected in 2012. The election featured eight candidates, but they were all from the only political party allowed, the Democratic Party of Turkmenistan.

Turkmenistan has huge deposits of metallic ores, including bentonite, bromine, gypsum, iodine, salt, sodium sulfate, sulfur, and zinc, as well as oil and natural gas reserves.

Carpet making is a major industry in Turkmenistan. The country's handmade woolen carpets, woven in brilliant hues against a deep background, are prized the world over for their color and design.

Crops in Turkmenistan require irrigation to grow. Cotton and wheat are the chief crops. Other farm products include beef, fruits and vegetables, milk, wool, and Persian lamb (a fur taken from young lambs). Some farmers also raise camels, horses, silkworms, and the famous Karakul sheep. Karakul sheep pelts are highly valued for fur coats.

The Karakum Canal carries water from the Amu Darya River across the desert to provide irrigation for crops. However, the canal diverts water from the Aral Sea. As a result, the water level of the Aral Sea has dropped steadily, creating serious ecological problems.

Turkmenistan is bordered by Afghanistan and Iran in the south, Kazakhstan and Uzbekistan in the east and northeast, and the Caspian Sea in the west.

TUVALU

A tiny island country in the South Pacific Ocean, Tuvalu has a land area of 10 square miles (26 square kilometers). Vatican City, Monaco, and Nauru—another Pacific island country—are the only nations in the world smaller than Tuvalu. The nine islands of Tuvalu are spread over about 360 miles (580 kilometers) and lie about 2,000 miles (3,200 kilometers) northeast of Australia.

Most of Tuvalu's islands are atolls, ring-shaped coral reefs that surround lagoons. People live on eight of the nine islands, giving the country its name, which means eight standing together. Funafuti, an islet, or small island, is the capital of Tuvalu. It has about 2,800 people.

Funafuti is one of the world's smallest and most unusual national capitals. It is the largest of 30 islets in an atoll that is also called Funafuti. The main government offices of Tuvalu, as well as a hospital, a hotel, and a jail, are on the islet. Funafuti also has a wharf and an airport.

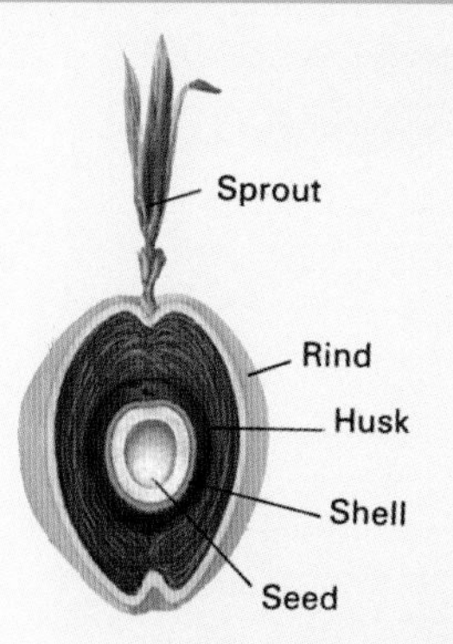

The coconut fruit has four layers: the seed, with its sweet, white meat; the seed's hard, brown shell; the husk of reddish-brown fibers; and a smooth, light-colored rind. Plants, seeds, and even animals are carried to otherwise lifeless coral islands by floating debris, birds, or the wind.

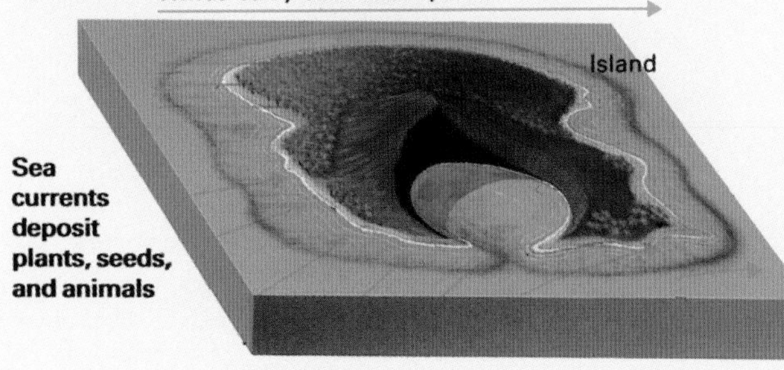

FACTS

Official name:	Tuvalu
Capital:	Funafuti
Terrain:	Very low-lying and narrow coral atolls
Area:	10 mi² (26 km²)
Climate:	Tropical; moderated by easterly trade winds (March to November), westerly gales and heavy rain (November to March)
Main rivers:	N/A
Highest elevation:	Unnamed location, 16 ft (5 m)
Lowest elevation:	Pacific Ocean, sea level
Form of government:	Constitutional monarchy with a parliamentary democracy
Head of state:	British monarch, represented by governor general
Head of government:	Prime minister
Administrative areas:	None
Legislature:	Fale I Fono (Parliament or House of Assembly) with 15 members serving four-year terms
Court system:	8 Island Courts, High Court
Armed forces:	N/A
National holiday:	Independence Day - October 1 (1978)
Estimated 2010 population:	12,000
Population density:	1,200 persons per mi² (462 per km²)
Population distribution:	51% rural, 49% urban
Life expectancy in years:	Male, 64; female, 68
Doctors per 1,000 people:	0.9
Birth rate per 1,000:	24
Death rate per 1,000:	8
Infant mortality:	27 deaths per 1,000 live births
Age structure:	0-14: 32%; 15-64: 62%; 65 and over: 6%
Internet users per 100 people:	41
Internet code:	.tv
Languages spoken:	Tuvaluan, English, Samoan, Kiribati
Religions:	Church of Tuvalu (Congregationalist) 97%, Seventh-Day Adventist 1.4%, other 1.6%
Currency:	Australian dollar
Gross domestic product (GDP) in 2008:	N/A
Real annual growth rate (2008):	N/A
GDP per capita (2008):	N/A
Goods exported:	Copra, fish
Goods imported:	Food, machinery, petroleum products
Trading partners:	Australia, China, Fiji, Japan

The first inhabitants of Tuvalu probably came from Samoa hundreds of years ago. In 1568, a Spanish explorer named Álvaro de Mendaña became the first European to see part of Tuvalu. However, the islands remained largely unknown to Europeans until the early 1800's. In that period, an American captain visited the island group and named them the Ellice Islands, in honor of his ship's owner. In the late 1800's, slave traders rounded up and sold islanders to work in plantations and mines elsewhere in the Pacific and in Latin America.

The United Kingdom took control of the island group in the 1890's. In 1916, the British combined the Ellice Islands with the Gilbert Islands to the north, forming the Gilbert and Ellice Islands Colony. In 1975, the two island groups were separated. The Gilbert Islands became the nation of Kiribati, and the Ellice Islands were renamed Tuvalu. The United Kingdom granted Tuvalu independence in 1978.

Tuvalu is a constitutional monarchy. It is a member of the Commonwealth of Nations, a group of countries that have lived under British law and government. A prime minister, chosen by a legislature elected by the people, heads the government.

Nanumea Atoll
Niutao Island
Nanumanga Island
North
Nui Atoll
Vaitupu Island
Nukufetau Atoll
Funafuti
Funafuti Atoll
Nukulaelae Atoll
South Pacific Ocean
Niulakita Island
0 100 200 Miles
0 100 200 300 Kilometers

The nation of Tuvalu consists of a group of nine islands, mostly atolls, spread over about 360 miles (580 kilometers). Niulakita, the southernmost island, is uninhabited. The islands were formerly known as the Ellice Islands.

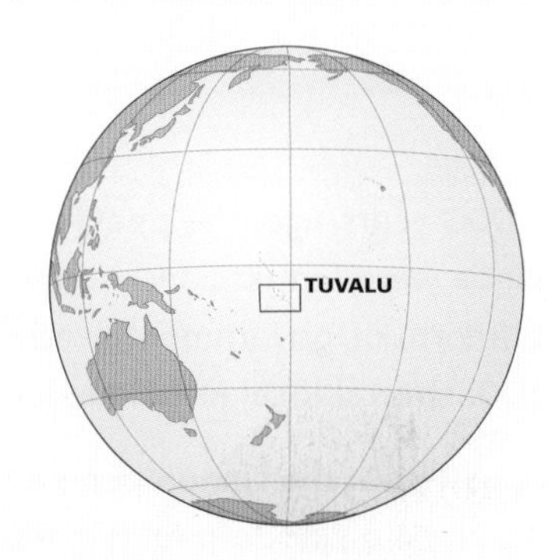

A traditional sing-song provides a social atmosphere for a group of Tuvaluans. Most of Tuvalu's people are Polynesians whose ancestors probably came from Samoa hundreds of years ago. The islands remained largely unknown to Europeans until the early 1800's.

Most of Tuvalu's 12,000 people are Polynesian and live in villages clustered around a church and a meeting house. The people speak the Tuvaluan language, and many also know English. The main foods of the Tuvaluans are bananas, coconuts, fish, and taro, a tropical plant with one or more edible rootlike stems. The islanders also raise pigs and chickens to eat at feasts.

Tuvalu has poor soil, few natural resources, and almost no manufacturing. Coconut palm trees cover much of the land, and the islanders use the coconuts to produce copra (coconut meat)—their chief export. Tuvalu also exports hand-woven baskets and mats. Many young islanders work on ocean ships because of the lack of jobs at home. Tuvalu receives aid from other countries, including Australia and the United Kingdom.

UGANDA

Uganda is a densely populated country in east-central Africa. Since the early 1970's, Uganda has been troubled by war and political instability.

Government

The president of Uganda, who is elected by the people for a five-year term, appoints a cabinet to help carry out the operations of the government. The people also elect a parliament, whose members serve five-year terms.

History

About 2,000 years ago, the people living in what is now Uganda were farmers and ironworkers. Later, they adopted a form of government headed by chiefs, and after 1300, several local kingdoms developed.

By about 1850, the Ganda people had formed the rich and powerful kingdom of Buganda, which had a large army and a highly developed government. During the 1860's, explorers and missionaries from the United Kingdom began arriving in the region. In 1894, Buganda was made a British protectorate that eventually included three other kingdoms.

An independence movement became strong among the Ganda people during the 1950's, and occasionally trouble erupted between the Kabaka (king) of Buganda and the British. Uganda became an independent nation on Oct. 9, 1962, but each of the four kingdoms that had been united in the protectorate kept its own king until 1967.

Milton Obote, a member of a northern ethnic group, became prime minister of Uganda in 1962. In 1963, Sir Edward Mutesa II, the Kabaka of Buganda, was elected president. When serious differences arose between the two, Obote dismissed Mutesa and took on the presidency himself. In 1967, a new constitution made Uganda a republic and abolished the kingdoms.

The government of Uganda has changed hands many times, at times being civilian and at other times military. Major General Idi Amin Dada ruled as a cruel dictator from 1971 until 1979. Along with his supporters, Amin ordered many thousands of Ugandans to be killed.

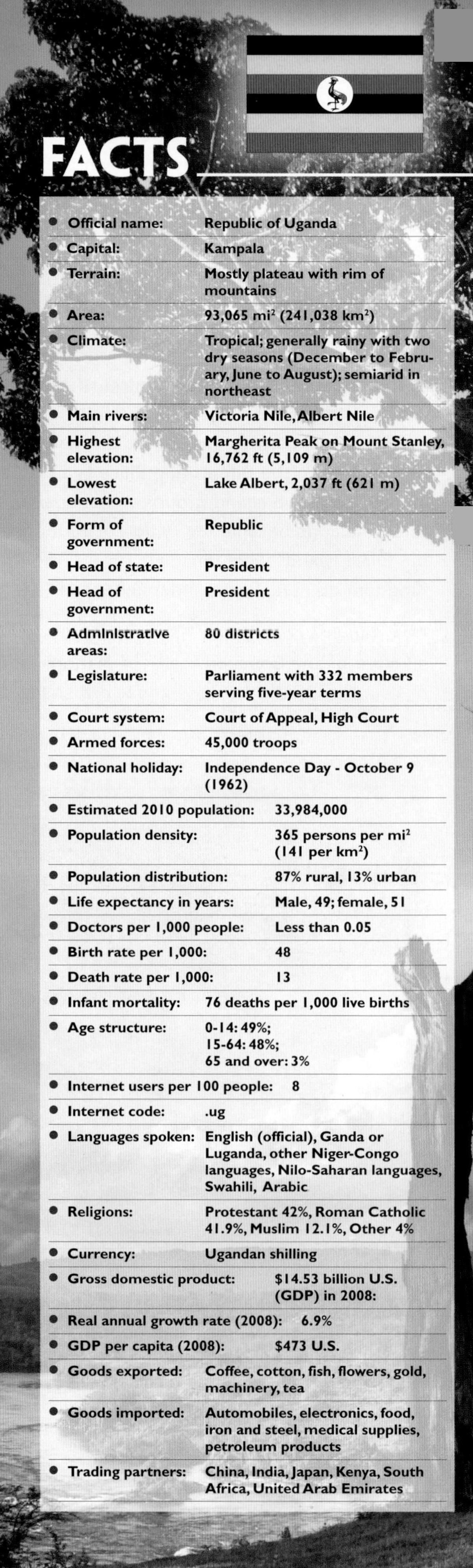

FACTS

Official name:	Republic of Uganda
Capital:	Kampala
Terrain:	Mostly plateau with rim of mountains
Area:	93,065 mi² (241,038 km²)
Climate:	Tropical; generally rainy with two dry seasons (December to February, June to August); semiarid in northeast
Main rivers:	Victoria Nile, Albert Nile
Highest elevation:	Margherita Peak on Mount Stanley, 16,762 ft (5,109 m)
Lowest elevation:	Lake Albert, 2,037 ft (621 m)
Form of government:	Republic
Head of state:	President
Head of government:	President
Administrative areas:	80 districts
Legislature:	Parliament with 332 members serving five-year terms
Court system:	Court of Appeal, High Court
Armed forces:	45,000 troops
National holiday:	Independence Day - October 9 (1962)
Estimated 2010 population:	33,984,000
Population density:	365 persons per mi² (141 per km²)
Population distribution:	87% rural, 13% urban
Life expectancy in years:	Male, 49; female, 51
Doctors per 1,000 people:	Less than 0.05
Birth rate per 1,000:	48
Death rate per 1,000:	13
Infant mortality:	76 deaths per 1,000 live births
Age structure:	0-14: 49%; 15-64: 48%; 65 and over: 3%
Internet users per 100 people:	8
Internet code:	.ug
Languages spoken:	English (official), Ganda or Luganda, other Niger-Congo languages, Nilo-Saharan languages, Swahili, Arabic
Religions:	Protestant 42%, Roman Catholic 41.9%, Muslim 12.1%, Other 4%
Currency:	Ugandan shilling
Gross domestic product:	$14.53 billion U.S. (GDP) in 2008:
Real annual growth rate (2008):	6.9%
GDP per capita (2008):	$473 U.S.
Goods exported:	Coffee, cotton, fish, flowers, gold, machinery, tea
Goods imported:	Automobiles, electronics, food, iron and steel, medical supplies, petroleum products
Trading partners:	China, India, Japan, Kenya, South Africa, United Arab Emirates

In July 1985, a group called the National Resistance Movement (NRM) began fighting to overthrow the government. They captured the capital city of Kampala in January 1986. The NRM leader, Yoweri Museveni, then became president.

From 1986 until 1994, a National Resistance Council (NRC) served as Uganda's legislature. In March 1994, a Constituent Assembly replaced the NRC. The Assembly approved a new constitution in 1995, and elections for a president and a parliament were held in 1996. Museveni was elected president and reelected in 2001. In 2005, Ugandans voted to restore multiparty democracy, which had been suspended since 1986. Also in 2005, Parliament amended the Constitution to remove term limits for the president. Museveni was reelected president in 2006 and 2011.

In 2001, Uganda, Kenya, and Tanzania launched the East African Community to promote regional economic and political cooperation. In 2007, Burundi and Rwanda joined the EAC.

In 2006, the Ugandan government and a rebel group called the Lord's Resistance Army (LRA), which had been fighting the government since the 1980's, signed a cease-fire. The LRA had carried out thousands of murders, rapes, and child kidnappings. The parties were to sign a peace agreement in 2008, but the LRA leader went into hiding and the group itself fled to neighboring Democratic Republic of the Congo. From there, the LRA continued to operate.

Worshipers gather outside a Roman Catholic church in Kampala, Uganda's capital. About 85 percent of Uganda's people are Christian, a heritage that goes back to the 1800's, when British missionaries and explorers came to the region.

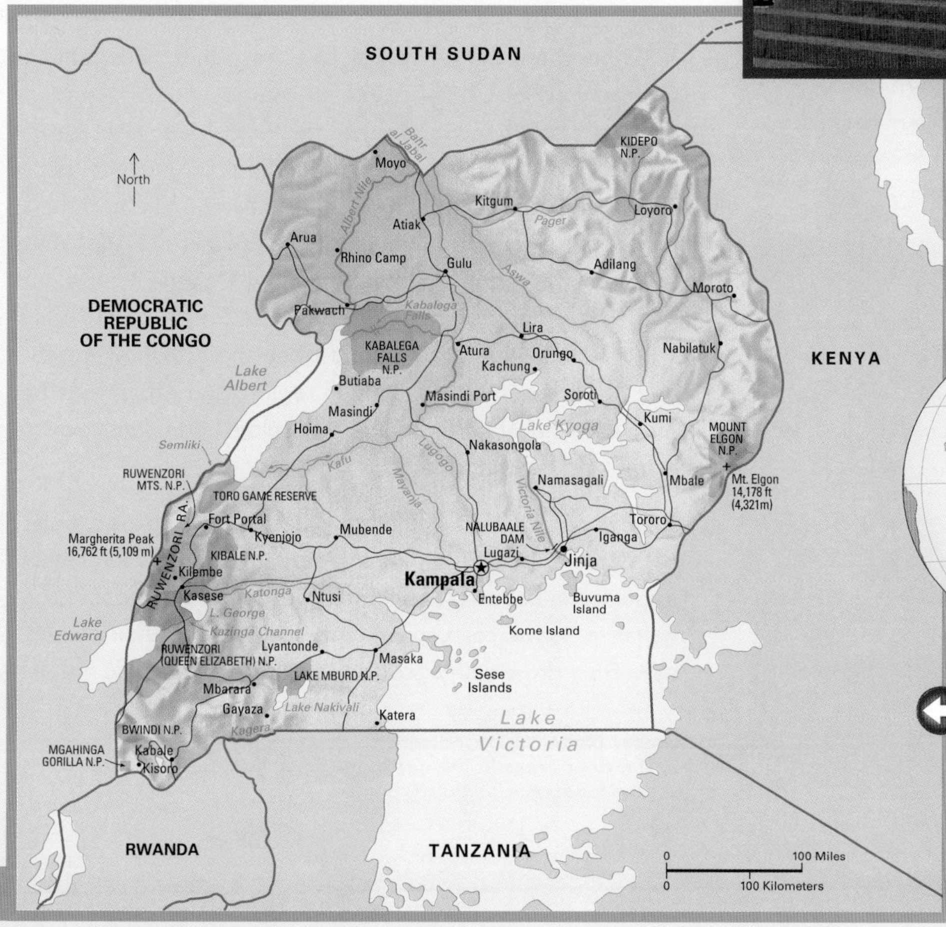

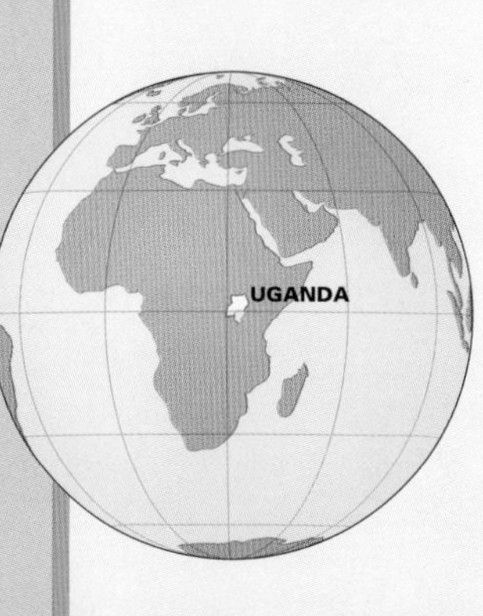

Uganda is a densely populated agricultural country in east-central Africa. Although the nation has rich mineral deposits, only copper is mined on a large scale. Many cargo ships operate on Uganda's large lakes.

ENVIRONMENT AND PEOPLE

Uganda has a magnificent landscape, ranging from towering, snow-capped mountains to dense tropical forests. Large lakes cover more than a sixth of the nation. A great variety of wild animals roam vast national parks such as Kabalega Falls National Park in the northwest.

Most of the country consists of a plateau about 4,000 feet (1,200 meters) above sea level. Thick rain forests grow in the south. In the north, grassy savannas stretch across most of the land, though some northeastern areas are near-deserts.

A Ugandan transports bananas to market on the back of a bicycle. Bananas, one of the chief food crops of Ugandans, are grown mainly for the people themselves rather than for export.

Highlands run along the eastern and western borders. Mount Elgon towers 14,178 feet (4,321 meters) near the border with Kenya. The Ruwenzori Range, which includes the 16,763-foot (5,109-meter) Margherita Peak, rises along the border with Congo (Kinshasa). Just east of these western mountains lies the Great Rift Valley, a series of cracks in Earth's surface.

Within the Great Rift Valley lie Lakes Albert, Edward, and George. Part of Lake Victoria—the largest lake in Africa and the world's second largest freshwater lake after Lake Superior in North America—forms the southeastern corner of Uganda. The Victoria Nile and the Albert Nile and their branches, which form part of the headwaters of the White Nile, drain most of Uganda. Several of Uganda's lakes and rivers were named for members of the British royal family. Victoria was queen of England when the British explored and began to rule the Ugandan region, and Albert was her husband and prince. Edward, their eldest son, later became king of England.

Although Uganda has many rich mineral deposits, only copper is mined to any great degree. Uganda is mainly an agricultural country, and its people grow such crops as bananas, cassava, corn, millet, and potatoes for their own use. They grow coffee, cotton, sugar cane, tea, and to-

Grassy plains and woodlands cover much of Uganda's land. In the drier areas in the north, the trees thin out, leaving a sparse landscape of thorny shrubs.

Kabalega Falls plunges into the Victoria Nile River between Lake Victoria and Lake Albert. A beautiful national park surrounding the waterfall is home to a variety of wildlife, including elephants, hippopotamuses, and leopards.

bacco mostly for export. Uganda's manufacturing industries process food products and also make cement and other construction materials, consumer goods, and textiles.

The African people of Uganda belong to more than 20 different ethnic groups. Almost every ethnic group has its own language, and no language is understood by everyone. English is the nation's official language.

The Ganda, the largest and wealthiest ethnic group in the country, live in the central and southern parts of Uganda. They speak a language called Ganda or Luganda, which belongs to the Bantu language group. Hundreds of years ago, ancestors of the Ganda developed the powerful kingdom of Buganda. Today, their political and social organization is one of the most highly developed in central Africa. Most Ganda are farmers, and Ganda women do much of the farm work. The people live in houses that have iron roofs and walls of cement, cinder block, or mud.

Most of Uganda's other ethnic groups are farmers as well. But the Karamojong in the northeast and several other groups in the drier sections of the country are herders who roam the land in search of pasture for their livestock.

About 85 percent of Uganda's people are Christians. Many others practice traditional African religions. About 10 percent are Muslims.

More than half of Uganda's adults can read and write. However, the wars and political unrest since the early 1970's have greatly reduced the educational opportunities for the nation's children.

A traditional dwelling makes a cozy sleeping place for Karamojong people of northeastern Uganda. The structures are made of clay, branches, and thatch.

Missionaries distribute food to children in northern Uganda. The people in this dry region face hardship when drought reduces the amount of pasture available to their cattle.

UKRAINE

Ukraine is a rich farming, industrial, and mining country in eastern Europe. It is bordered by the Black Sea to the south and the Carpathian Mountains to the west. Ukraine consists mostly of fertile steppes (vast plains).

Ukraine is one of the world's leading farming regions, and it is sometimes called the breadbasket of Europe. The country also manufactures iron and steel, tractors, machine tools, transportation equipment, and many other products. About a fourth of Ukraine's people work in industry, and about a fifth work in agriculture. Most other Ukrainians have jobs in such service industries as education and health care.

History

The long history of the Ukrainian people began with prehistoric agricultural tribes who inhabited the Dnieper and Dniester river valleys. During the A.D. 800's, a Slavic civilization developed along the river routes between the Black Sea and the Baltic Sea. The region around Kiev, called Kievan Rus, became the first East Slavic state. Vladimir I, ruler of the Russian principality of Novgorod, conquered Kievan Rus in 980. He made Christianity the state religion in 988.

In 1240, Mongol tribes known as Tatars swept across the plains from central Asia and conquered Ukraine. They soon left, and the region broke up into small states. In the mid-1300's, some of the area was taken over by the expanding Lithuanian and Polish states. Russia began moving into the region in the 1600's and gained control over most of Ukraine by 1795.

In 1918, after the Bolsheviks (later called Communists) came to power in Russia, the Ukrainians established an independent non-Communist state. But by 1920, Russian Communists brought most of Ukraine under their rule. In 1922, Ukraine became one of the four original republics of the Soviet Union.

Nazi Germany occupied Ukraine from 1941 to 1944, during World War II. After the war, the Soviet Union reestablished its control. The Ukrainian Insurgent Army fought first German and then Soviet control into the early 1950's. The Ukrainians' opposition to Soviet domination and its restrictions on their cultural

FACTS

Official name:	Ukrayina (Ukraine)
Capital:	Kyyiv (Kiev)
Terrain:	Most of Ukraine consists of fertile plains (steppes) and plateaus, mountains being found only in the west (the Carpathians), and in the Crimean Peninsula in the extreme south
Area:	233,090 mi^2 (603,700 km^2)
Climate:	Temperate continental; Mediterranean only on the southern Crimean coast; precipitation highest in west and north, lesser in east and southeast; winters vary from cool along the Black Sea to cold farther inland; summers are warm across the greater part of the country, hot in the south
Main rivers:	Dnieper, Dniester, Donets
Highest elevation:	Mount Hoverla, 6,762 ft (2,061 m)
Lowest elevation:	Black Sea, sea level
Form of government:	Republic
Head of state:	President
Head of government:	Prime minister
Administrative areas:	24 oblasti, 1 avtomnaya respublika republic), 2 mista (municipalities) with oblast status
Legislature:	Verkhovna Rada (Supreme Council) with 450 members serving five-year terms
Court system:	Supreme Court, Constitutional Court
Armed forces:	129,900 troops
National holiday:	Independence Day - August 24 (1991)
Estimated 2010 population:	45,378,000
Population density:	195 persons per mi^2 (75 per km^2)
Population distribution:	68% urban, 32% rural
Life expectancy in years:	Male, 62; female, 74
Doctors per 1,000 people:	3.1
Birth rate per 1,000:	10
Death rate per 1,000:	16
Infant mortality:	11 deaths per 1,000 live births
Age structure:	0-14: 14%; 15-64: 70%; 65 and over: 16%
Internet users per 100 people:	23
Internet code:	.ua
Languages spoken:	Ukrainian (official), Russian, Romanian, Polish, Hungarian
Religions:	Ukrainian Orthodox - Kiev Patriarchate 50.4%, Ukrainian Orthodox - Moscow Patriarchate 26.1%, Ukrainian Greek Catholic 8%, Ukrainian Autocephalous Orthodox 7.2%, other 8.3%
Currency:	Hryvnia
Gross domestic product (GDP) in 2008:	$180.36 billion U.S.
Real annual growth rate (2008):	2.1%
GDP per capita (2008):	$3,916 U.S.
Goods exported:	Chemicals, food, fuels, iron and steel, machinery
Goods imported:	Chemicals, machinery, natural gas, oil, vehicles, wood products
Trading partners:	China, Germany, Italy, Poland, Russia, Turkey, Turkmenistan

Ukrainian farms produce sugar beets, meat and dairy products, and grain. Ukrainian farmers also raise barley, cabbages, corn, potatoes, sunflowers, tobacco, tomatoes, and other crops.

freedom continued through the 1970's and 1980's. Nevertheless, Ukraine remained under the strict control of the Soviet central government.

In 1991, in the midst of political upheaval in the Soviet Union, Ukraine declared its independence. The Soviet Union was formally dissolved in December. After becoming independent, Ukraine shifted to an economy based on the free market and private ownership of business. From 1991 to 2001, the government converted farms owned or managed by the state into privately owned farms.

In early 1992, tension developed between Russia and Ukraine over control of the former Soviet naval fleet in the Black Sea. The countries reached an agreement on how to divide the fleet in 1997. In 1994, Ukraine signed an agreement with the United States and Russia to transfer its long-range nuclear weapons to Russia, where they would be destroyed. This transfer was completed in 1996. Also in 1996, Ukraine adopted a new constitution.

In 2004, Ukraine underwent a shift toward democratic government. A disputed presidential election led to massive peaceful protests called the Orange Revolution. Ukraine's Supreme Court called for a new election, and Viktor Yushchenko, the democratic reformer candidate, won. In 2006, his rival, Viktor Yanukovych, won parliamentary elections and became prime minister under Yushchenko. Yanukovych won the 2010 presidential election.

People

About four-fifths of the people in Ukraine are Ukrainians, a Slavic nationality group. Russians, who make up about one-sixth of the population, are the next largest group. Most Ukrainians belong to Eastern Orthodox churches. Other religious groups include Ukrainian Catholics, Protestants, Jews, and Muslims. The official language is Ukrainian, which has several local dialects.

Ukraine is Europe's second largest country in area. Most of the country has cold winters and warm summers.

THE UNITED ARAB EMIRATES

The United Arab Emirates (UAE) is a federation of seven independent Arab states at the southern end of the Persian Gulf. From west to east, the states are Abu Dhabi, Dubai, Ash Shariqah, Ajman, Umm al Qaywayn, Ras al Khaymah, and Al Fujayrah. The capital city of each state has the same name as the state. Abu Dhabi is the capital of the federation.

Each of the seven states of the UAE is an *emirate*, ruled by a prince called an *emir*. The emir controls his own state's political and economic affairs.

The seven emirs form the Supreme Council of Rulers. The council elects a president to serve as chief executive. The president appoints a prime minister to head a Council of Ministers, which supervises government departments. The emirs appoint 20 representatives to a legislature called the Federal National Council. Voters chosen by each emir elect another 20 members. The federal government handles foreign affairs, defense, and economic and social development.

The Arab states that now make up the UAE began to develop during the 1700's, when European nations had already established trading posts in the Persian Gulf area. The British eventually became the strongest foreign power in the region.

Ras al Khaymah and Ash Shariqah became the first strong Arab states in the area. Their strength came from naval power as well as from the wealth obtained from pearl diving and trading.

In the late 1700's and early 1800's, Ras al Khaymah and Ash Shariqah warred with other gulf states over control of the region's trade. Not only did the British aid the rivals of the two Arab states, but they destroyed the city of Ras al Khaymah.

In 1820, the British forced all the states to sign a truce forbidding warfare at sea. Other truces were later signed, and the region became known as the Trucial States. But rivalries over boundaries, pearl-diving rights, and other disputes continued into the mid-1900's.

FACTS

Official name:	**Al Imarat al Arabiyah al Muttahidah (United Arab Emirates)**
Capital:	**Abu Dhabi**
Terrain:	**Flat, barren coastal plain merging into rolling sand dunes of vast desert wasteland; mountains in east**
Area:	**32,278 mi² (83,600 km²)**
Climate:	**Desert; cooler in eastern mountains**
Main rivers:	**N/A**
Highest elevation:	**Jabal Yibir, 5,010 ft (1,527 m)**
Lowest elevation:	**Salamiyah, a salt flat slightly below sea level**
Form of government:	**Federation of emirates (territories ruled by emir)**
Head of state:	**President**
Head of government:	**Prime minister**
Administrative areas:	**7 imarat (emirates)**
Legislature:	**Majlis al-Ittihad al-Watani (Federal National Council) with 40 members serving two-year terms**
Court system:	**Union Supreme Court**
Armed forces:	**51,000 troops**
National holiday:	**Independence Day - December 2 (1971)**
Estimated 2010 population:	**4,765,000**
Population density:	**148 persons per mi² (57 per km²)**
Population distribution:	**78% urban, 22% rural**
Life expectancy in years:	**Male, 75; female, 79**
Doctors per 1,000 people:	**1.7**
Birth rate per 1,000:	**16**
Death rate per 1,000:	**2**
Infant mortality:	**7 deaths per 1,000 live births**
Age structure:	**0-14: 20%; 15-64: 79%; 65 and over: 1%**
Internet users per 100 people:	**65**
Internet code:	**.ae**
Languages spoken:	**Arabic (official), Persian, English, Hindi, Urdu**
Religions:	**Muslim 96%, other (including Christian and Hindu) 4%**
Currency:	**Emirati dirham**
Gross domestic product (GDP) in 2008:	**$260.14 billion U.S.**
Real annual growth rate (2008):	**7.7%**
GDP per capita (2008):	**$55,068 U.S.**
Goods exported:	**Crude oil, dates, fish, natural gas**
Goods imported:	**Chemicals, food, machinery, transportation equipment**
Trading partners:	**China, Germany, India, Japan, South Korea, United States**

Dubai is the largest city in the United Arab Emirates. The city's skyline is dominated by the Burj Khalifa. Completed in 2010, the 2,712-foot (828-meter) tower is the world's tallest building.

By that time, Abu Dhabi and Dubai had become the leading states in the area. The United Kingdom controlled all the states' foreign affairs, while guaranteeing them protection against invaders. Each emir continued to handle his own state's internal matters.

Then in the mid-1900's, foreign oil companies began to drill for oil in the Trucial States. In 1958, petroleum was discovered in Abu Dhabi, and in 1966 large oil deposits were found in Dubai.

In 1971, the Trucial States gained full independence from the United Kingdom. Despite their long-time feuding, six of the states decided to form a union called the United Arab Emirates. The seventh—Ras al Khaymah—joined the federation in 1972.

Oil production began in Ash Shariqah in 1974, and oil profits enabled Abu Dhabi, Dubai, and Ash Shariqah to develop into modern states. However, the oil industry brought people as well as wealth to the UAE. Thousands of workers came from neighboring Arab countries—as well as from Bangladesh, India, Iran, Pakistan, and the Philippines—to work in the oil industry.

In 1981, the UAE and other states of eastern Arabia formed the Gulf Cooperation Council (GCC) to work together in such matters as defense and economic projects. In early 1991, the UAE and the other GCC members took part in the allied air and ground offensive that liberated Kuwait from the Iraqis who had invaded Kuwait during the previous year.

The citizens of the UAE are Arab Muslims from tribes that have lived in the region for hundreds of years. However, most of the people living in the UAE are not citizens. They are people who have come there from other countries to work. Nearly 80 percent of the inhabitants live in urban areas, mainly in modern houses or apartments. Most citizens prefer traditional Arab clothing, but many immigrants wear clothing of their country of origin. The official language is Arabic.

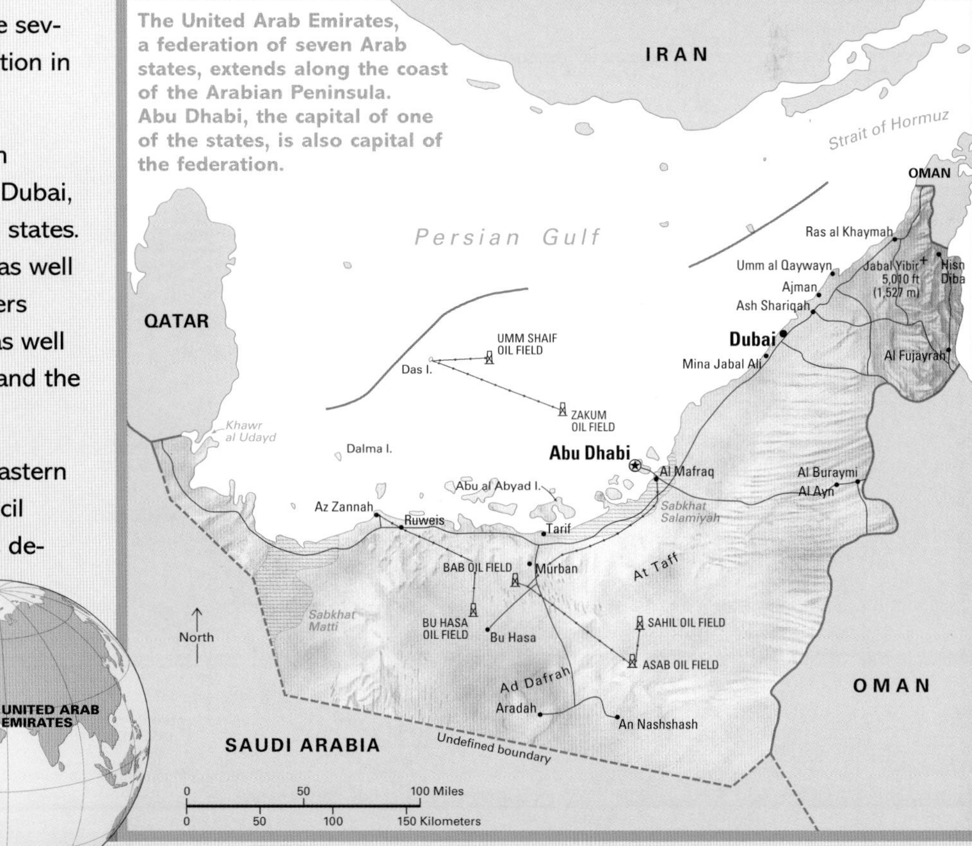

The United Arab Emirates, a federation of seven Arab states, extends along the coast of the Arabian Peninsula. Abu Dhabi, the capital of one of the states, is also capital of the federation.

LAND AND ECONOMY

Before the mid-1900's, the region that is now the United Arab Emirates (UAE) was one of the most underdeveloped areas in the world.

The land itself has few natural resources, and the climate is hot with little rainfall. Humid swamps and salt marshes line much of the northern coast, while desert covers most of the inland area. Hills and mountains rise up in the eastern section.

Until the 1950's, most of the UAE's people earned a living by diving for pearls, fishing, herding camels, growing dates, or trading. The discovery of oil, beginning in the 1950's, brought sudden wealth into the region. New industries were created, and modern cities began to develop. Many of the people left their traditional ways of life and took jobs in the oil industry or other modern occupations.

Under the Provisional Constitution adopted in 1971, the emirs of the seven states of the UAE agreed to share their resources and work for the economic development of all the states. During the decade of the 1970's, the UAE's average income per person was among the highest in the world.

The Mall of the Emirates in Dubai is one of the largest shopping malls in the world. It includes more than 350 stores as well as the Middle East's first indoor ski resort, a theatre, an indoor amusement park, a multiplex cinema, and a hotel.

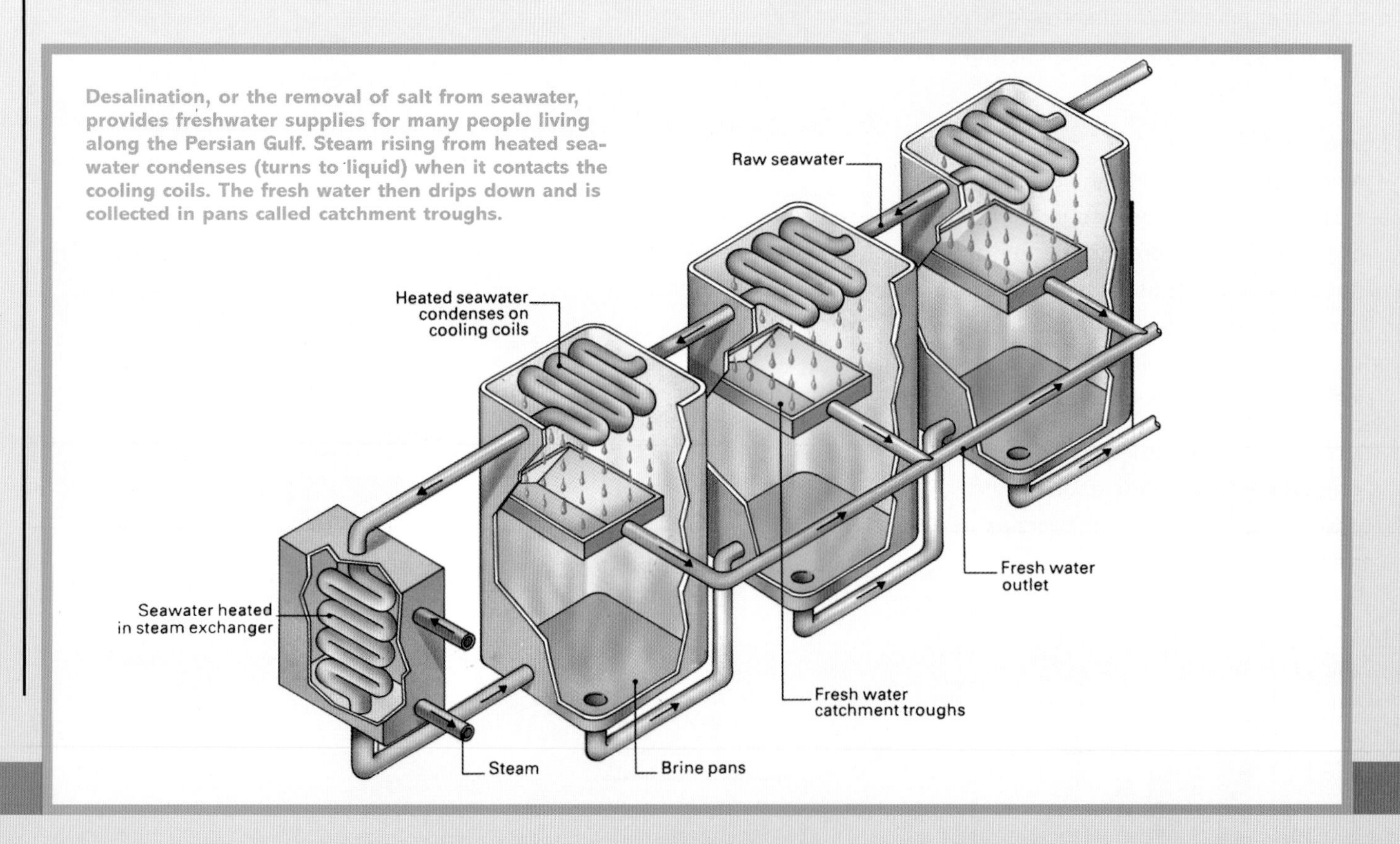

Desalination, or the removal of salt from seawater, provides freshwater supplies for many people living along the Persian Gulf. Steam rising from heated seawater condenses (turns to liquid) when it contacts the cooling coils. The fresh water then drips down and is collected in pans called catchment troughs.

Arab musicians, keeping traditions alive, hold the attention of a group of children. Oil income has enabled the United Arab Emirates to build many schools. Today, most UAE children receive a good education, and about 90 percent of all adults can read and write.

DUBAI

Of all the United Arab Emirates, Dubai has made the greatest effort to attract foreign businesses and investment, especially from Europe and North America. In the early 1970's, the state began an ambitious program to develop an international airport and harbor. Port Rashid is now a major shipping center, where goods are transferred from oceangoing vessels to dhows (Arabian sailing ships). Some of the dhows carry goods back and forth to the coast of Iran, less than 100 miles (160 kilometers) away. Persian carpets are just one of the many items traded in this duty-free zone.

An oceangoing container ship awaits unloading at one of the berths in the modern shipping center at Port Rashid in Dubai.

When thousands of people came from neighboring countries to find jobs, the UAE was faced with housing shortages and other problems. But profits from the oil industry and other economic activities have helped the federal government build apartments, schools, hospitals, and roads to meet the needs of the growing population.

Most of the oil production of the UAE takes place in the states of Abu Dhabi, Dubai, and Ash Shariqah. The rulers of these emirates earn huge profits selling oil to foreign countries. The exports are mainly crude oil, but the UAE also has refineries that process some of the oil into petroleum products.

The production of natural gas, which often is found near petroleum, also brings income into the UAE. In addition, the region has trading and banking facilities.

Other states of the UAE have also begun producing some oil. However, they continue to rely on agriculture and fishing as the basis of their economies.

Desert covers much of the region, so less than 1 percent of the land of the UAE can be farmed. However, wells and oases dot the desert, and the hills and mountains in the eastern section generally receive more rainfall than the rest of the area. Farmers in the desert oases and the highlands grow dates, melons, tomatoes, and other crops. Desert nomads tend herds of camels, goats, and sheep.

People who live in the coastal areas earn their living by fishing. The UAE exports small amounts of fish and dates. Dubai, Abu Dhabi, and Ash Shariqah are the main ports.

THE UNITED KINGDOM

The United Kingdom is a country in northwestern Europe. It consists of four political divisions—England, Scotland, and Wales, which make up the island of Great Britain, and Northern Ireland, which occupies the northeastern part of the island of Ireland.

The nation's official name is the United Kingdom of Great Britain and Northern Ireland. When people refer to the country, most shorten its name to the United Kingdom, the U.K., or Britain, and they call its people British. Although all British people share certain customs and traditions, each division has its own dialect, culture, history, and traditions. Most of the people have a strong sense of regional identity.

The United Kingdom is slightly smaller than the state of Wyoming and has about 1 percent of the world's total population. Yet for hundreds of years, it was one of the world's most important countries. The British started the Industrial Revolution and founded the largest empire in history. For centuries, they ruled the seas, and their nation ranked as the world's greatest trading nation.

By the end of World War II (1939-1945), however, the power of the United Kingdom had been much reduced. The crippling costs of two world wars and competition from other industrial countries led to one economic crisis after another. By the early 1950's, the United Kingdom's empire was declining rapidly, as many of its former colonies became independent nations.

Today, the United Kingdom remains a leading industrial and trading nation, but it is no longer the world power it once was. Most of the nations that were once part of the empire are now linked with the United Kingdom and with one another through the Commonwealth of Nations. This association of free countries and other political units recognizes the British monarch as head of the Commonwealth. However, the monarch is mainly symbolic and has no governing power.

Although the United Kingdom is part of Europe, it is separated from mainland Europe by the North Sea on the east and by the English Channel on the south. Throughout history, this separation has protected the lands from invasion and helped shape the independent character of the British people. But in May 1994, a 31-mile (50-kilometer) railway beneath the English Channel opened. The Channel Tunnel, nicknamed "the chunnel," is an outstanding engineering feat that allows passengers to travel by rail from London to Paris in about 3 hours.

THE UNITED KINGDOM TODAY

The United Kindom is a constitutional monarchy. The British Constitution is not a single document. It consists partly of *statutes* (laws passed by Parliament) and of documents such as the *Magna Carta* (a charter passed in 1215 to limit the monarch's power). It also includes *common law* (laws based on custom and supported in the courts). Much of the Constitution is not even written. These unwritten parts include many important ideas and practices that have developed over the years.

Number 10 Downing Street, guarded night and day by the London police, has been the official residence of the United Kingdom's prime ministers since 1732.

Government

The British monarch, the United Kingdom's official head of state, holds mainly ceremonial responsibilities. A Cabinet of government officials, called *ministers*, actually rules the country. The Cabinet is responsible to Parliament, which makes the laws of the United Kingdom.

Parliament consists of the monarch, the House of Commons, and the House of Lords. The monarch must approve all bills passed by Parliament before they become laws, but no monarch has rejected a bill since the early 1700's.

The prime minister, who is usually the leader of the political party with the most seats in the House of Commons, serves as the head of government. The monarch appoints the prime minister after each general election. The prime minister selects the Cabinet ministers. In 1999, the central government turned over some of its powers to the governments of Scotland, Wales, and Northern Ireland.

People

The United Kingdom is a densely populated country, and about 90 percent of the people live in urban areas. English is the official language, but some people in Wales, Scotland, and Northern Ireland speak the traditional language of their areas.

FACTS

Official name:	United Kingdom of Great Britain and Northern Ireland
Capital:	London
Terrain:	Mostly rugged hills and low mountains; level to rolling plains in east and southeast
Area:	93,628 mi² (242,495 km²)
Climate:	Temperate; moderated by prevailing southwest winds over the North Atlantic Current; more than one-half of the days are overcast
Main rivers:	Thames, Severn, Mersey, Humber, Clyde
Highest elevation:	Ben Nevis, 4,406 ft (1,343 m)
Lowest elevation:	Holme Fen, 9 ft (2.7 m) below sea level
Form of government:	Constitutional monarchy; in practice, a parliamentary democracy
Head of state:	Monarch
Head of government:	Prime minister
Administrative areas:	England: 34 two-tier counties, 32 London boroughs and 1 City of London or Greater London, 36 metropolitan counties, 46 unitary authorities Northern Ireland: 26 district councils Scotland: 32 unitary authorities Wales: 22 unitary authorities
Legislature:	Parliament consisting of House of Lords with about 750 members and House of Commons with 646 members serving five-year terms
Court system:	House of Lords
Armed forces:	160,300 troops
National holiday:	N/A
Estimated 2010 population:	61,489,000
Population density:	657 persons per mi² (254 per km²)
Population distribution:	90% urban, 10% rural
Life expectancy in years:	Male, 77; female, 81
Doctors per 1,000 people:	2.3
Birth rate per 1,000:	13
Death rate per 1,000:	9
Infant mortality:	5 deaths per 1,000 live births
Age structure:	0-14: 18%; 15-64: 66%; 65 and over: 16%
Internet users per 100 people:	80
Internet code:	.uk
Languages spoken:	English, Welsh, Scottish
Religions:	Christian 71.6%, Muslim 2.7%, other 25.7%
Currency:	British pound
Gross domestic product (GDP) in 2008:	$2.674 trillion U.S.
Real annual growth rate (2008):	0.7%
GDP per capita (2008):	$44,134 U.S.
Goods exported:	Aerospace equipment, chemicals and pharmaceuticals, foods and beverages, machinery, motor vehicles, petroleum
Goods imported:	Clothing, food, machinery, metals, motor vehicles, petroleum products, textiles
Trading partners:	Belgium, France, Germany, Netherlands, Spain, United States

The United Kingdom covers most of an island group called the British Isles. The British Isles consist of Great Britain, Ireland, and many small islands.

About a third of the United Kingdom's urban residents live in England's seven metropolitan areas. Other densely populated regions include Scotland's Central Lowlands, southern Wales, and the Belfast area of Northern Ireland.

Most of the British are descendants of the many early peoples who came to the islands, including the Celts, Romans, Angles, Saxons, Jutes, Danes, and Normans. But since World War II ended in 1945, many immigrants from Commonwealth countries in the West Indies, Asia, and Africa have settled in the United Kingdom. The nation has also offered asylum to refugees from around the world.

There are many divisions in British life. For example, Scotland and England have their own national churches. There are also separate legal and educational systems in England and Wales, Scotland, and Northern Ireland. For centuries, the British people were also separated by a rigid class system, but these class barriers were greatly reduced during and after World War II.

ENGLAND

England, in the southern and eastern part of the island of Great Britain, ranks as the largest of the four political divisions that make up the United Kingdom. With more than 80 percent of the total British population, England is a densely populated area. About 95 percent of its people live in urban areas. But outside the crowded city centers stretches the scenic English countryside, with its charming villages, green pastures, and neat hedges.

Land

In general, England's land slopes from the north and west to the south and east. Characteristic features include *moors* (open grasslands), *downs* (hilly grasslands), *fens* (marshlands), and *wolds* (low, chalky hills).

England's Lake District where 15 lakes lie within a circle about 30 miles (48 kilometers) in diameter, is celebrated for its beauty.

The Pennines—England's major mountain system—extend from Scotland to central England. They are often called the "backbone of England." West of the Pennines lies the Lake District, known for its beautiful mountain scenery, including England's highest point—the 3,210-foot (978-meter) Scafell Pike.

The rugged Southwest Peninsula, located south of Wales, consists of a low plateau and highlands. It includes England's westernmost point—Land's End—and the southernmost point in the British Isles—Lizard Point.

The English lowlands cover the rest of England and are home to most of its farmland, industry, and people. A large plain called the Midlands occupies the center of the lowlands and holds England's chief industrial cities. Farther south, the land is broken by rolling hills and fertile valleys. Along the North Sea coast, the land is low and flat, particularly in the Fens, where much of the land has been reclaimed from the sea.

South of the Thames River, ranges of hills consisting of layers of limestone and chalk cross the land. Along the English Channel, the hills drop sharply to form steep cliffs, including the famous white cliffs of Dover.

England's offshore islands include the Isle of Wight, off the southern coast, and the Scilly Islands, off Land's End.

The Tyne Bridge spans the Tyne River in Newcastle upon Tyne. The city is a commercial and manufacturing center of northern England.

Blenheim Palace, the seat of the Duke of Marlborough, is in Woodstock, Oxfordshire. England's largest country house, Blenheim was built by Queen Anne for the first Duke of Marlborough as a reward for winning the Battle of Blenheim in 1704. World War II Prime Minister Sir Winston Churchill, grandson of the seventh duke, was born at Blenheim on Nov. 30, 1874.

Elegant townhouses, built in the 19th century, line a quiet street in fashionable Belgravia, a residential area in central London.

Economy

Until the early 1800's, most English people lived in the countryside and worked on farms. Then, during the Industrial Revolution, huge numbers of people moved to cities and towns to work in the new factories.

Today, more than a third of England's population lives in seven large metropolitan areas. Greater London, with more than 7 million people, is the largest metropolitan area in England and one of the largest in the world. England's six metropolitan counties, with the largest city of each in parentheses, are Greater Manchester (Manchester), Merseyside (Liverpool), South Yorkshire (Sheffield), Tyne and Wear (Newcastle upon Tyne), West Midlands (Birmingham), and West Yorkshire (Leeds).

The shift in population from rural to urban areas reflected England's shift from an agricultural to an industrial economy. In recent years, however, many of the factories built near coal fields, their source of power, have closed. Nuclear energy, oil, and gas are the modern energy sources. As a result, many new industries have developed around London and in the southeastern section of England, where there is little coal. The decline of the factories around the northern coal fields has led to a drop in that region's prosperity and to unemployment problems. The new industries in the south have drawn even more people to an already crowded area.

Service industries are important to England's economy. About 70 percent of English workers are employed in service industries, particularly in social services, wholesale and retail trade, and financial services. Other service industries are concerned with communication, education, and leisure activities. Popular leisure activities include gardening, sports, watching television, and attending movies, plays, and concerts.

England has a long and rich cultural heritage, with contributions from numerous important writers, musicians, artists, and craftspeople. Today, London is a world center for music and drama. Birmingham and other major English cities also have a number of music and theater companies. Two of the most famous universities in the world, Oxford and Cambridge, are in England. The University of London is England's largest traditional university.

WALES

Wales covers about 10 percent of the island of Great Britain, but it has only about 5 percent of the population of the United Kingdom. About one-sixth of the people speak Welsh, an ancient Celtic language. English and Welsh are both official languages.

Wales has a history filled with poets and singers, with a literature that dates back more than 1,000 years and an ancient choral music tradition. A festival called the eisteddfod (pronounced ay STEHTH vahd), featuring musicians, poets, and singers, is held once a year.

The Pass of Llanberis lies southeast of Llanberis. Llanberis is the starting point for the rail ascent of Snowdon, the highest peak in Wales. Snowdonia National Park is nearby.

Government changes

As part of the United Kingdom, Wales elects 40 members of the House of Commons. In addition, in 1997, Welsh voters approved a plan to form a National Assembly of Wales. The Assembly, which first met in 1999, has 60 elected members. Although the Assembly has no tax-raising powers, it has considerable powers within Wales. Its responsibilities include economic development, education, health, transportation, and culture. At first, the Assembly only had the power to pass secondary legislation, which decided the details of how to administer certain primary legislation passed by the Parliament of the United Kingdom. In 2007, the National Assembly gained the authority to pass primary legislation for Wales in some specific cases.

Land and economy

The Cambrian Mountains cover about two-thirds of Wales. The highest peak in Wales, Snowdon (Yr Wyddfa in Welsh), reaches 3,561 feet (1,085 meters). Coastal plains and river valleys cover about a third of Wales.

The longest rivers are the Severn and Wye, which both empty into the Bristol Channel. The Isle of Anglesey (Ynus Môn in Welsh) is a large island off the northwest coast of Wales. It is separated from the mainland by the Menai Strait. Wales has three scenic national parks—Snowdonia, the Brecon Beacons, and the Pembrokeshire Coast. All three are popular with hikers and climbers.

Caernarfon Castle is one of the most famous castles in Wales. It was begun in 1283 and completed in the early 1300's. The castle stands on the Seiont River in the town of Caernarfon in northwest Wales.

The Royal National Eisteddfod, a festival of music and the arts, takes place in August. It is held in various cities and towns, alternately in southern and northern Wales. Only Welsh is officially used during this event.

About four-fifths of the Welsh people live in urban areas, mostly in the industrial regions of southern and northeastern Wales. The largest cities are Cardiff, Swansea, and Newport, all ports are located on the southern coast. Cardiff has been the capital since 1955, and its importance has grown since the establishment of the National Assembly.

A major increase in Wales's population occurred during the Industrial Revolution in the 1700's, when large numbers of people immigrated from England to work in the coal mines. Many Welsh people also left their farms to find work in the coal-mining towns of southern Wales, where most mining activities were concentrated.

By the late 1900's, however, only a few mines were still operating. The economy now depends primarily on manufacturing and service industries. Light industry, such as the production of electronic equipment, has become increasingly important.

Most farms are small, and livestock raising is important. Farmers raise beef and dairy cattle in the lowlands and sheep in the uplands.

The Pierhead Building with its famous clock tower is a Cardiff landmark. The building was completed in 1897 as the headquarters of a shipping company. It became a historical museum in 2010. The Wales Millennium Center stands in the background with the Welsh National Assembly building on the right.

SCOTLAND

Scotland occupies the northern third of the island of Great Britain, but only about 9 percent of the total population of the United Kingdom lives there. Most of Scotland's population is concentrated in an industrial area located in the central part of Scotland.

The Scots have a strong sense of identity, and they have maintained numerous symbols of their long and colorful history. Still, industrialization has altered many of Scotland's old traditions. The historic Scottish *clans* (groups of related families) have lost much of their importance, and kilts are now usually worn only on special occasions.

A bagpiper in traditional Highland costume plays at a gathering of the clans. The bagpipe, which dates back thousands of years, is one of the oldest musical instruments still in use.

Scots participate in the government of the United Kingdom and elect representatives to the British House of Commons. But since 1999, they also have a 129-member Scottish Parliament. The Scottish Parliament has power to impose some taxes and to control education, health, the environment, agriculture, and the arts in Scotland. In addition, Scotland's legal system is distinct from that of the rest of the United Kingdom. The Scottish legal system was preserved by the 1707 Treaty of Union, which joined England, Wales, and Scotland under one government.

Loch Awe is one of the many beautiful lakes in the Scottish Highlands. On its banks stand the ruins of Kilchurn Castle, built in about 1440 and once the scene of fierce battles between clans.

Land

Scotland has three main land regions: the Highlands, the Central Lowlands, and the Southern Uplands. The Highlands are a magnificent, rugged region that covers the northern two-thirds of Scotland. The area's two major mountain ranges, the Northwest Highlands and the Grampian Mountains, are divided by a deep valley called *Glen Mor* (Great Glen). The Highlands are dotted with sparkling *lochs*, or lakes.

The Central Lowlands, an area of scenic green valleys, fertile fields, and scattered woodlands, has Scotland's

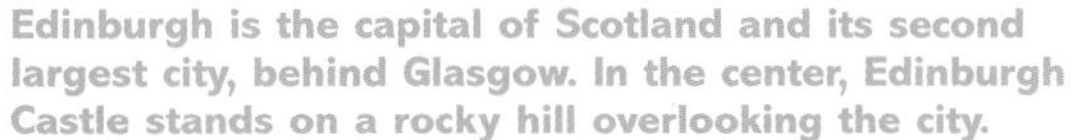

Edinburgh is the capital of Scotland and its second largest city, behind Glasgow. In the center, Edinburgh Castle stands on a rocky hill overlooking the city.

A distillery at Tain on Dornoch Firth, along with Scotland's many other distilleries, exports huge quantities of Scotch whisky each year. The biggest market is the United States.

Mallaig, on the northwest coast of Scotland, a is an important fishing port. Scotland is renowned for its salmon and its successful development of a salmon farming industry.

richest farmland, most of its mineral resources, and about three-fourths of its people. The region includes two of Scotland's leading cities—Glasgow, its largest city, in the west, and Edinburgh, its capital, in the east.

The Southern Uplands consist of rolling moors, broken in places by rocky cliffs. Sheep and cattle are raised on the rich pastureland that covers most of the lower slopes.

Economy

Most of the Scottish people live in industrial cities and towns. After the decline of heavy manufacturing in the late 1900's, service industries became the main employers of Scotland's work force. More than two-thirds of Scotland's people work in such areas as retail sales, finance and business services, tourism, transportation, and communication.

Oil and gas fields under the North Sea provide much of Scotland's energy. Among Scotland's most valuable resources are the fishing areas off the east and north coasts. Chemicals, electronic equipment, metal products, refined petroleum, steel, textiles, and whisky also contribute to Scotland's economy.

About two-thirds of Scotland's land is used for farming, but less than one-fifth of the land is suitable for crops. Farmers use most of the land for grazing cattle and sheep. The animals are raised for milk, meat, and wool.

In their leisure time, the Scots enjoy sports, including golf, which is believed to have originated in Scotland, and soccer. Highland Games, held throughout the Highlands during the spring, summer, and early fall, include field events, foot races, and dancing and bagpipe competitions.

NORTHERN IRELAND

Northern Ireland consists of the northeastern section of the island of Ireland. It is often called *Ulster*.

Ulster was the name of a large province of British-controlled Ireland until 1920. In 1920, the United Kingdom separated Northern Ireland from the rest of Ireland in order to create separate governments for the predominantly Protestant north and the mostly Roman Catholic south. The majority of the Northern Irish people supported the separation, but many Roman Catholics in both the north and the south refused to accept the division. In 1921, the south became the self-governing Irish Free State, now the independent Republic of Ireland.

Beginning in 1921, militant Irish groups, particularly the Irish Republican Army (IRA), attacked British government installations in Northern Ireland, hoping to force the British to give up control. Protestant groups retaliated. The conflict, based on long-standing political and economic disputes, carried on for decades. In the 1960's, the Roman Catholic minority held marches demanding an end to discrimination. The police reacted violently, and riots broke out. In 1968, the IRA and Protestant paramilitaries began committing acts of violence. The United Kingdom sent troops in 1969 and established direct rule over Northern Ireland in 1972.

On April 10, 1998, formal peace talks led to the Belfast Agreement (also called the Good Friday Agreement), which committed all sides to using peaceful means to resolve political differences. The talks also established a power-sharing legislature called the Northern Ireland Assembly.

The agreement brought greater stability to Northern Ireland. However, some unrest caused delays in establishing the new government bodies that had been proposed. The Northern Ireland Assembly met in 2002 but then was suspended until 2007. Also in 2007, the British army officially ended its military campaign, which had lasted since 1969. In 2010, the transfer of responsibility for police and justice issues from British authorities to Northern Ireland took place. The transfer was the last transfer of power called for by the Good Friday Agreement.

Land

Northern Ireland is a land of rolling plains and low mountains. The fertile plains cover the central part of the area, and scenic green valleys and low mountains lie along the coast.

The countryside of Northern Ireland is dotted with smooth, clear lakes called *loughs* (pronounced *lahks*). Lough Neagh, the largest lake in the British Isles, covers

150 square miles (388 square kilometers) near the center of Northern Ireland.

The Lusty Man—a stone carving of an ancient god dating from about A.D. 500—greets visitors coming and going at Lough Erne in the district of Fermanagh. The statue has two faces, carved on opposite sides.

A farm woman prepares a picnic lunch for harvest workers near the Mourne Mountains in the district of Down.

Economy

About 15 percent of the people of Northern Ireland live in Belfast, the capital and largest city. Belfast is also Northern Ireland's manufacturing and trading center. Many of Northern Ireland's linen mills, shipyards, and aircraft plants are located there.

Northern Ireland's main manufacturing areas involve engineering, communications, aerospace, pharmaceuticals, and textiles. However, the economy depends mainly on service industries, including education, e-commerce, government services, health care, telecommunications, and tourism.

Since fertile pastureland is Northern Ireland's chief natural resource, agriculture is also an important industry. About 20 percent of Northern Ireland's people live in rural areas, but few of them make their living by farming. Northern Ireland's fishing fleet processes cod, herring, mackerel, shrimp, and whiting, mainly from the Irish Sea.

Life in Northern Ireland is much like life in the rest of the United Kingdom. Such sports as soccer, cricket, and golf are popular.

An aircraft plant in Belfast manufactures many aerospace products and ranks as a major employer in Northern Ireland.

Royal Avenue is one of the major shopping areas in Belfast. The street is lined with fashionable shops and hotels.

CHANNEL ISLANDS AND ISLE OF MAN

The Channel Islands are a group of islands in the English Channel. Although they lie only about 10 to 30 miles (16 to 50 kilometers) off the coast of France, the islands have been attached to the English Crown since the 1000's.

The Isle of Man lies in the Irish Sea midway between England and Ireland. Great Britain has controlled the Isle of Man since 1765. But British laws do not apply to the island unless it is specifically named in the legislation.

The Channel Islands

The four main islands of the Channel Islands are Jersey, Guernsey, Alderney, and Sark. Along with numerous smaller islets, the islands cover 75 square miles (194 square kilometers). The total population of the Channel Islands is about 157,000. English is the official language, but many islanders speak a French dialect that varies from island to island.

The islands, which have been largely self-governing since the 1200's, are divided into two administrative units called *bailiwicks*. A lieutenant governor assigned to each bailiwick represents the British monarch and handles international affairs. The four main islands—Jersey, Guernsey, Alderney, and Sark—have their own parliaments to regulate internal affairs.

Several of the islands have their own distinguishing characteristics. For example, Jersey, the largest of the islands, is known for its cows and for its sweaters, which are often called "jerseys." In the 1600's, so many men abandoned their farmwork to knit the jerseys that a law was introduced to ban knitting in the summer months.

Alderney and Guernsey are also known for their cattle. The Guernsey breed of cattle, known for their rich milk, originated on the island.

A worker harvests potatoes on Jersey, the largest of the Channel Islands. Farmers on the island grow early potatoes, tomatoes, and other vegetables.

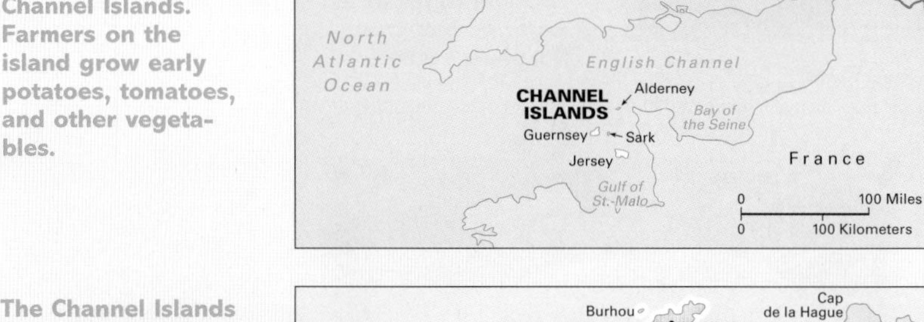

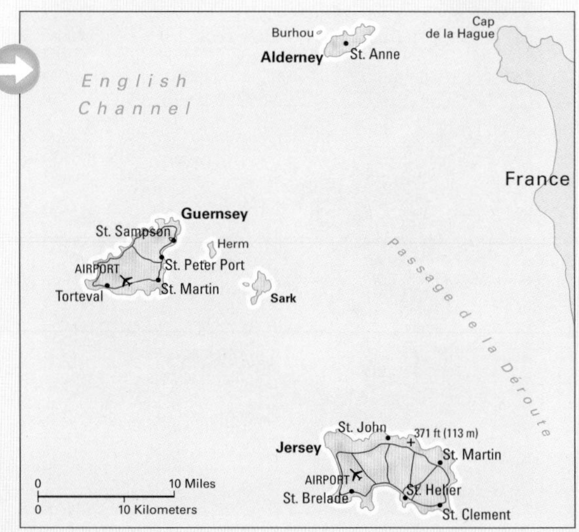

The Channel Islands lie in the English Channel. There are four main islands--Jersey, Guernsey, Alderney, and Sark. The group also includes several smaller islands and a number of tiny, rocky isles.

ISLE OF MAN

A dependency of the British Crown, the Isle of Man is an island in the Irish Sea, about halfway between England and Ireland and about 20 miles (32 kilometers) south of Scotland. The island has an area of 221 square miles (572 square kilometers) and a population of about 78,000. The people speak English, though some also speak a Celtic language called *Manx*.

A British lieutenant governor represents the monarch on the island. A more than 1,000-year-old parliament called Tynwald Court regulates the island's concerns.
Crowds of tourists visit summer resorts on the Isle of Man. Its international motorcycle race, held each June, draws many enthusiasts. Financial services form the island's leading economic activity, thanks to laws and tax policies that encourage companies from around the world to register their headquarters on the island. Other important industries include agriculture, fishing, and some high-technology businesses.

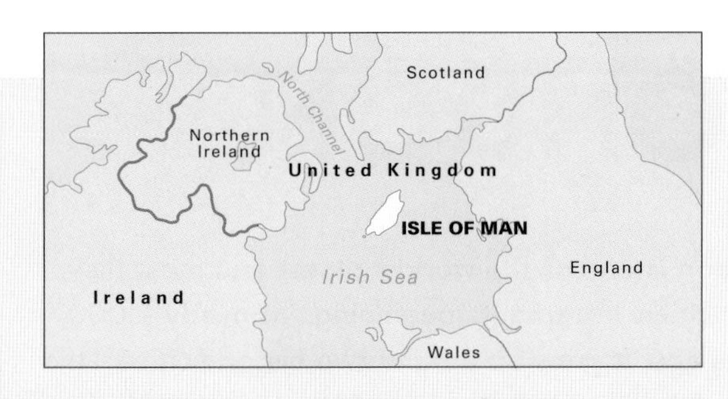

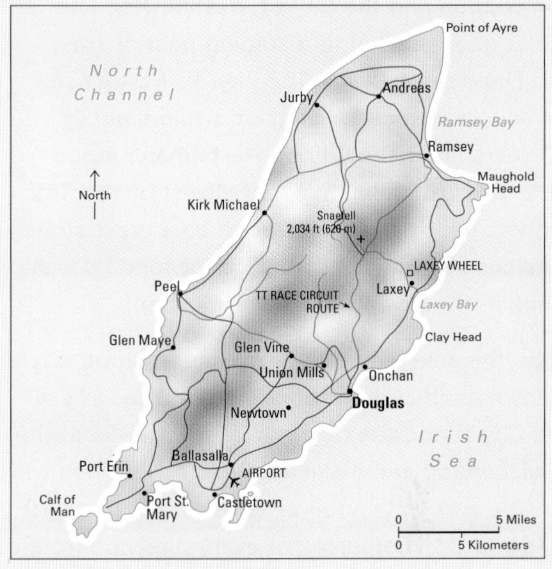

The Tynwald Ceremony annually celebrates the Isle of Man's system of self-government. The island's parliament, Tynwald Court, is older than England's.

A flower-covered float is on parade during Jersey's annual Battle of the Flowers, which takes place in July or August in St. Helier.

Sark, the smallest self-governing unit in the United Kingdom, has a democratic form of government headed by a *seigneur*, or feudal lord. The use of cars is prohibited on Sark, and the people travel by horse-drawn carriage, bicycle, or tractor. The island is only 3 miles (4.8 kilometers) long and 1-1/2 miles (2.4 kilometers) wide.

The leading industries in the Channel Islands are financial services and tourism. Financial services and banking have grown in importance since the 1990's. Hundreds of thousands of tourists visit each year, many drawn by the islands' pleasant beaches and historic landmarks. The mild climate and fertile soil help make farming important as well. Farmers grow fruits, vegetables, and flowers and raise beef and dairy cattle. There is little manufacturing on the islands.

LONDON

London is one of the world's oldest and most historic cities. It traces its beginnings to nearly 2,000 years ago. It grew up around two historic cities—the City of London and the City of Westminster. The City of London started as a trading post of the Roman Empire in A.D. 43. The City of Westminster began as a residence for England's rulers about 1,000 years later. The City of Westminster stood about 2 miles (3 kilometers) southwest of the City of London, which was surrounded by a great stone wall. As London grew, it spread far beyond its wall and took in the royal City of Westminster.

Today, the area where Roman London stood is still known as the City of London, often called simply the City. It and the City of Westminster lie at the heart of London and make up much of its lively center. The rest of London extends about 10 to 19 miles (16 to 31 kilometers) in every direction from this central section.

The City is London's financial district, and it consists largely of modern bank and office buildings. But it also contains such historic buildings as St. Paul's Cathedral, built by the great English architect Sir Christopher Wren between 1675 and 1710.

London architecture is a mix of old and new. The bulbous, glass Swiss Re Building, popularly called "the Gherkin," rises 590 feet (180 meters) over "The City," London's financial district. The low stone building along the River Thames is the Tower of London, a castle that dates from the 1000's.

The Trooping of the Colour in a parade ground known as the Horse Guards Parade is one of London's most spectacular ceremonies. The royal Household Cavalry also changes the guard daily at Horse Guards Parade.

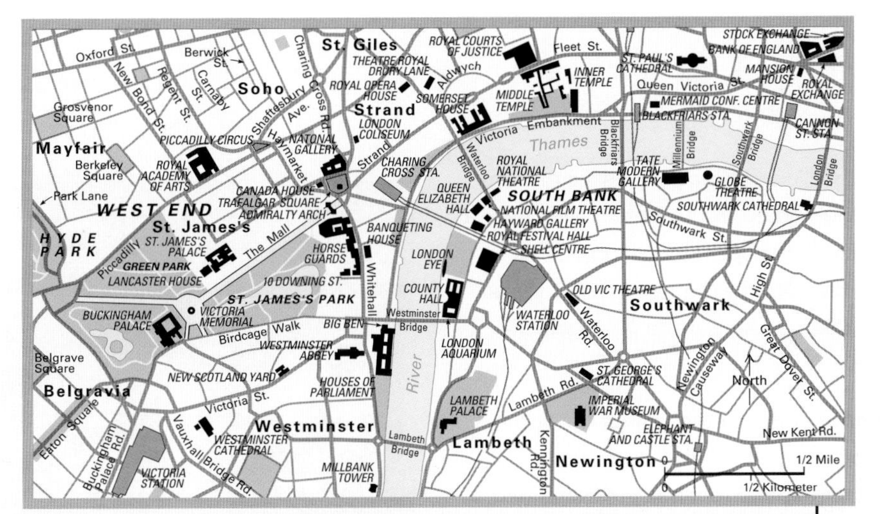

Greater London spreads across the River Thames Basin, extending outward from central London. Central London has tall office buildings and busy streets, as well as outstanding art galleries, museums, theaters, and beautiful parks. Its famous landmarks include Buckingham Palace, the Houses of Parliament, and St. Paul's Cathedral.

The Tower of London, which borders the City, dates from the late 1000's. It served as a royal fortress, palace, and prison and is now a national monument and museum. It houses the British crown jewels.

In Westminster, the center of the United Kingdom's government, are the Houses of Parliament, Westminster Abbey, and Buckingham Palace. Big Ben, the clock tower of the Houses of Parliament, is world famous. London's main shopping and entertainment districts lie nearby.

London has a number of professional theaters and world-renowned symphony orchestras. Numerous art galleries and museums are found in the city, and the South Bank section of central London is the site of a large, modern cultural center.

The Portobello Road Market is one of London's best-known street markets. Open on Saturdays, the market is known for its second-hand clothes and antiques.

The dome of St. Paul's Cathedral is one of London's most familiar landmarks. In the foreground, the Millennium Bridge crosses the River Thames from the Tate Modern, an art museum in Southwark.

A MULTIETHNIC SOCIETY

Most of the people of the United Kingdom descend from various early peoples who settled in the British Isles over a period of many centuries. Celtic-speaking people lived there by the mid-600's B.C. During the next 1,700 years, Romans, Angles, Jutes, Saxons, Danes, and Normans invaded the land.

Many other people have come to the United Kingdom seeking religious freedom or better living conditions. During the late 1600's, for example, French Protestants fled to the British Isles to escape persecution in their own country. Many Jews from Eastern Europe also sought refuge in the United Kingdom during the late 1800's and early 1900's.

Since the end of World War II in 1945, millions of immigrants have moved to the United Kingdom. Many have come from the Commonwealth of Nations, an association of independent nations and other political units that were once part of the British Empire. Most of these immigrants came from India, the West Indies, Pakistan, and Africa. Others arrived from Bangladesh, Hong Kong, and some Arab countries. More recently, the United Kingdom has offered asylum to many refugees from around the world.

A school playground in inner London reflects the multiracial nature of most large cities of the United Kingdom. The concentration of immigrants in many of the country's inner cities has resulted in crowded living conditions.

Initially, many of these immigrants came seeking factory jobs and settled in such industrial cities as London, Birmingham, and Leicester. Some immigrants also found employment in transportation and health services, while others established their own small businesses. Today, most British cities reflect the influence of immigrant cultures in many ways, including food. For example, Indian, Chinese, and other ethnic restaurants are as widespread as the more traditional British fish-and-chips shops.

The large numbers of immigrants crowding into British cities have resulted in a number of problems. Racial tension is particularly a problem in inner cities, where the inhabitants must often deal with poor living conditions and high unemployment rates. In these

People of many different backgrounds shop at an open-air market on a busy London street. They include Muslims from Bangladesh, immigrants from the Caribbean islands, and native Londoners.

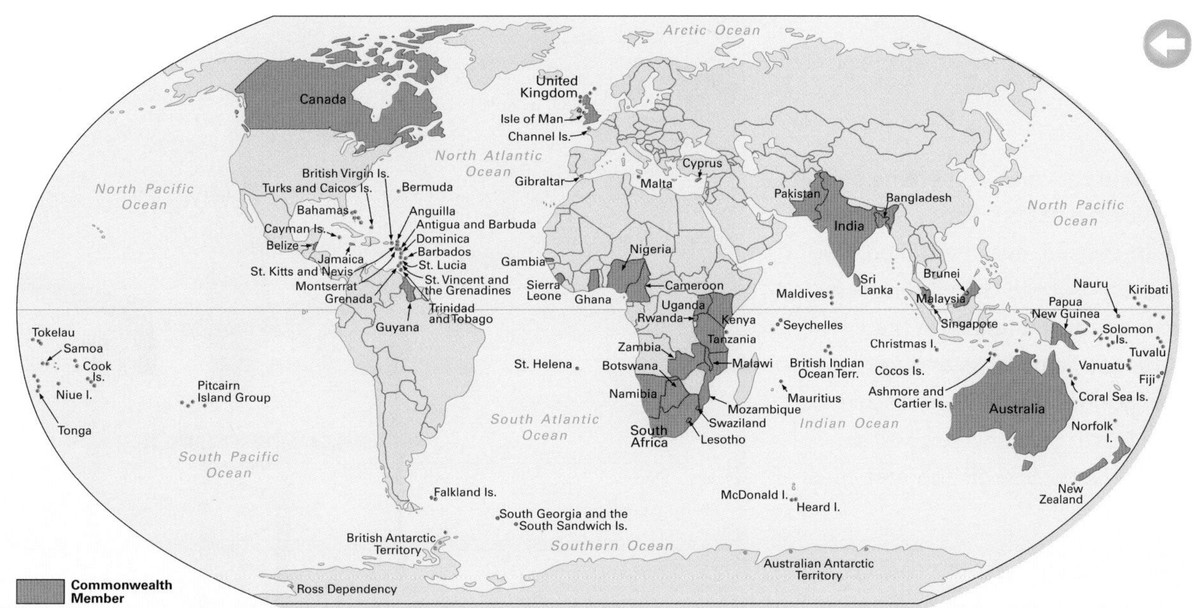

The Commonwealth of Nations includes the United Kingdom and many former British colonies. Many people have immigrated to the United Kingdom from independent Commonwealth countries and also from Commonwealth dependencies.

areas, many British people resent having to compete for jobs with the immigrants. At the same time, the immigrants resent the discrimination that they sometimes encounter. In 1981, this situation exploded in race riots in over 50 British cities and towns. Riots again broke out in urban areas in 1985.

Despite these problems, many people in the United Kingdom accept the society's multiracial structure. In some schools, the languages and histories of ethnic minorities are taught to encourage an appreciation of the country's different cultural heritages

Nonetheless, many ethnic groups fear that traditional ties of family loyalty and patriotism will be weakened by exposure to British culture. While many immigrants accept British citizenship, they also wish to preserve their cultural and religious identity. Thus, for example, they may continue to worship at Hindu temples or Islamic mosques. In addition, the ethnic peoples in many large British cities publish their own newspapers. They also celebrate their traditional festivals, such as Diwali (the Hindu Festival of Lights) and the Chinese New Year.

Throughout history, invaders and immigrants have been absorbed by the country and have become part of British society. In the process, they have added new elements to the richness of the nation's language and culture. Some people believe that the United Kingdom's situation today is part of the continuing interaction between new arrivals and established residents. Such interaction, they believe, will continue to expand and enrich the culture of the United Kingdom in the years to come.

Sikh women in London celebrate the festival of Vaisakhi, the beginning of the harvest season in their religion. Many Sikh immigrants have settled in England from India since the mid-1900's.

ECONOMY

The United Kingdom's plentiful coal and iron ore helped to make it the world's first industrial nation. Today, the demand for coal has greatly decreased, and the United Kingdom's iron ore is low in quality. But the United Kingdom has uncovered new sources of raw materials. In the 1960's and 1970's, natural gas fields large enough to supply all of the United Kingdom's needs were discovered in the southern part of the North Sea, and large oil deposits were discovered in the North Sea off the east coast of Scotland.

Early British factories were located near the coal fields, their source of power. But today, power from petroleum, natural gas, and nuclear energy has enabled many new industries to develop in other areas.

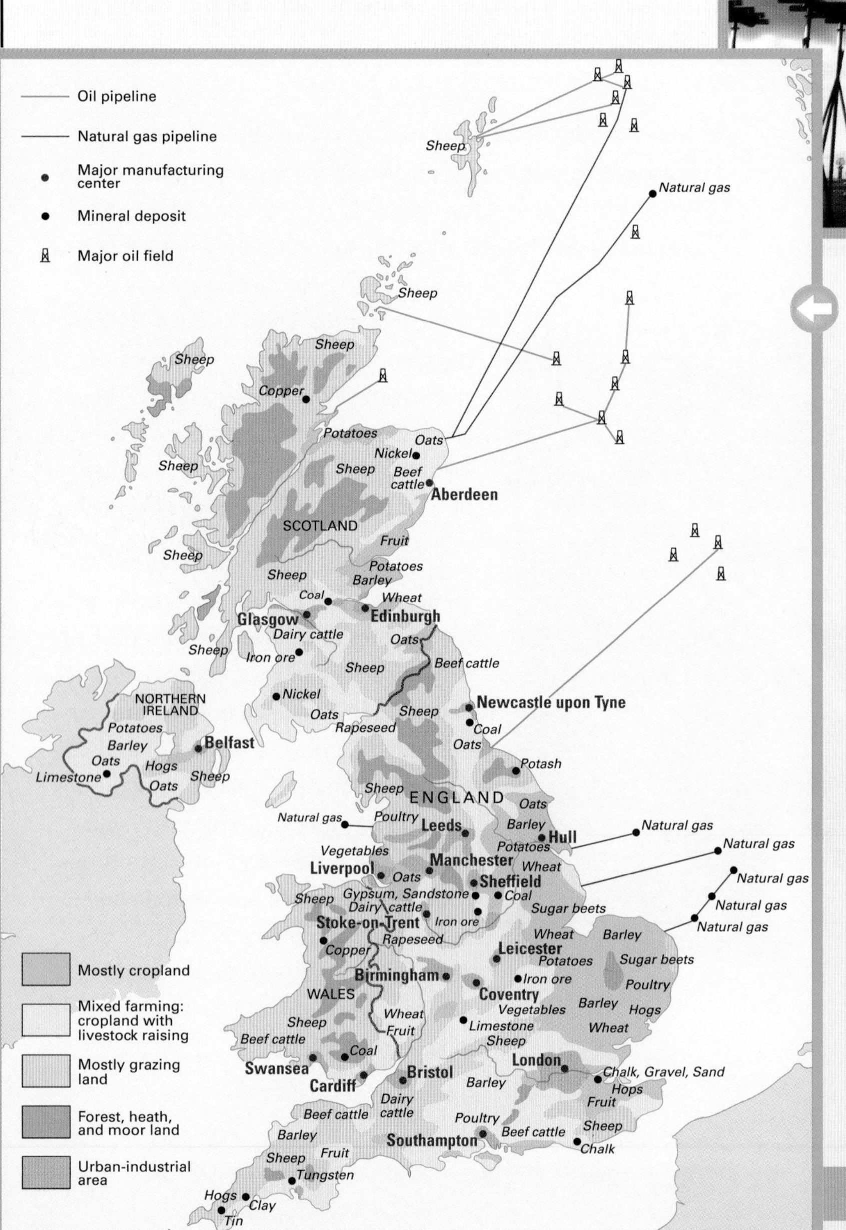

In the early 1900's, many British industries were concentrated in the rich coal-mining areas of Wales and Scotland. Today, the use of such materials as petroleum and natural gas has led to the growth of industry in other areas, particularly in southern England.

Industry and agriculture

The United Kingdom is a leading producer of manufactured goods. British factories have long been known for their production of automobiles, ships, steel, and textiles. Today, in addition to these goods, the country produces heavy machinery, aerospace equipment, electronics, and chemicals. Other major industries include printing and publishing, the production of clothing, and food and beverage processing. Most major British industries are in central England, the London area, the Scottish Central Lowlands, the Newcastle upon Tyne area, and southern Wales.

Service industries account for about 75 percent of the total value of goods and services produced within the country annually. About 80 percent of the United Kingdom's workers are employed in service industries. These industries include community, social, and per-

A jumbo jet prepares to land at Heathrow Airport outside London. Heathrow is one of the world's busiest airports, flying millions of people in and out of the country every year.

London is one of the world's leading fnancial cities. Most of the country's financial companies operate out of London.

sonal services; wholesale and retail trade; and finance, insurance, real estate, and business services.

Only about a third of British land is fertile enough for raising crops, but highly mechanized farming methods help the United Kingdom produce most of the food it needs. The country's most important crops are barley, potatoes, rapeseed (used to make livestock feed, vegetable oil, and industrial lubricants), sugar beets, and wheat. Farmers also raise beef and dairy cattle, sheep, hogs, and chickens.

Foreign trade and transportation

The United Kingdom ranks as a leading trading nation. It exports aerospace equipment, chemicals and medicines, foods and beverages, machinery, motor vehicles, petroleum, and scientific and medical equipment. Its imports include chemicals, clothing, food and beverages, machinery, metal ores, motor vehicles, paper and newsprint, petroleum products, and textiles. Most of the United Kingdom's trade is with other industrialized countries. The United Kingdom has an excellent network of roads and railroads and a large merchant fleet.

The value of British imports usually exceeds the value of its exports. To make up part of the difference, British banks and insurance companies sell their services to people and firms in other countries. Spending by the millions of tourists who visit each year is another important source of income.

PARTY POLITICS

Margaret Thatcher, prime minister from 1979 to 1990, worked to reduce government involvement in the economy and to encourage private enterprise.

The Labour Party, which campaigned on a socialistic program, came to power in 1945 and converted the United Kingdom to a welfare state. The new government expanded the social security system and launched a program of free medical care. The Labour government also began to put private industries, such as coal mining, steel, and the nation's railroads, under state control.

The measures were taken in an effort to restore the economy, which had suffered greatly during World War II. By 1955, economic conditions had improved.

In 1979, the Conservative Party, under Prime Minister Margaret Thatcher, began to reduce taxes and government involvement in the economy. The government began to sell state-owned industries to private buyers. The goal was to replace the welfare state with an "enterprise economy."

The Conservative government's policies and a decline in manufacturing eventually led to high unemployment, which peaked in 1986. Thatcher's policies grew unpopular, and she resigned in November 1990. John Major—chancellor of the exchequer in Thatcher's government—became prime minister. He sought a closer union with the Economic Community, a group of countries that formed the European Union in 1993.

A Labour Party victory in 1997 made Tony Blair prime minister. He supported efforts to broaden the appeal of Labour beyond the working class. Party members voted to drop a statement of socialist principles from the party constitution, and the influence of labor unions on the party was reduced. Gordon Brown, also of the Labour Party, succeeded Blair in 2007.

In 2008, the United Kingdom slid into a recession brought about by a global financial crisis. Brown's popularity, as well as that of the Labour Party, plummeted. In 2010, David Cameron, a Conservative, was elected prime minister. The country entered a second recession in 2012—the United Kingdom's first double-dip recession since the 1970's.

David Cameron became prime minister of the United Kingdom in 2010. Cameron, leader of the Conservative Party, replaced Prime Minister and Labour Party leader Gordon Brown.

FESTIVALS

An extraordinary variety of events is celebrated in the United Kingdom. Some festivals, such as the *eisteddfods* (festivals featuring poetry and music) in Wales and the Highland Games in Scotland, highlight national cultures. Others, such as the Trooping of the Colour, honor the reigning monarch.

Many regional festivals, such as the Up-Helly-A' festival in the Shetland Islands, recall aspects of an area's past. In addition, there are music festivals, theater festivals, religious festivals, and flower shows, as well as celebrations that accompany special horse races and sporting events.

Sports events in the United Kingdom draw many spectators. Thousands of fans crowd into stadiums to watch soccer, the nation's favorite sport. Rugby football, cricket, tennis, and golf are also popular. The United Kingdom participates in international matches in all these sports.

Festive themes

Many British festivals appeal to particular groups, such as the upper class, the working people, or members of other cultures who have immigrated to the country. For example, high society goes on parade at the Royal Ascot Races, particularly on Ladies Day. Much of the spectacle of this annual event is provided by the exquisitely dressed women and top-hatted men.

On the other side of the social coin, English miners unite to celebrate their profession in the Durham Miners' Gala. And in London, the Notting Hill Carnival celebrates the rich musical tradition of immigrants from the West Indies. The carnival, which began as an effort to soothe the cultural clashes that erupted soon after the immigrants arrived, is a joyous street party featuring parades, Caribbean music, dancing, and food.

The Notting Hill Carnival takes place on the streets of Notting Hill in London each August. It is a celebration of Caribbean culture by people who emigrated from Trinidad and Tobago beginning in the 1950's.

A morris dancer awaits his turn to enter the folk dance at a Whitsuntide festival in Exeter, England. Morris dancers often represent characters from Robin Hood stories.

The Henley Royal Regatta is a spectacular sporting and social event that takes place each summer. Parties of elegantly dressed women and men sporting straw hats watch the races while picnicking along the banks of the river. The rowing competitions were first held at Henley-on-Thames in 1839.

Spring and summer festivities

Many festivals celebrate the coming of spring. Whitsuntide, for example, a religious holiday observed throughout the United Kingdom on the seventh Sunday after Easter, celebrates returning hope and faith.

In May, the British, who are traditionally considered to be enthusiastic gardeners, turn out for the Chelsea Flower Show in London. The Glyndebourne opera season also begins in May with pomp.

Summertime brings more flower shows, horse shows, and—particularly in Scotland and Wales—many fascinating sheepdog trials. This is also the time for huge, tourist-attracting festivals, such as the Edinburgh International Festival held from mid-August to early September.

It is impossible to list all the festivals that take place in the United Kingdom because, it seems, there is something to celebrate on almost every day of the year.

CALENDAR OF EVENTS

A sampler of British festivals, by location

Ascot: Royal Ascot Races (June)

Badminton: Horse Trials (April or May)

Braemar: Braemar Royal Highland Gathering (September)

Belfast: Belfast Festival at Queen's, an arts festival (autumn)

Brighton: Arts Festival (May)

Cambridge: Festival of Nine Lessons and Carols, a Christmas Eve service at King's College (December)

Epsom: The Derby (June)

Edinburgh: Edinburgh Festival (August-September)

Farnborough: Farnborough Air Show (July)

Glastonbury: Glastonbury Fair (August)

Glyndebourne: Opera season (opens May)

Henley: Royal Regatta (June-July)

Padstow: Padstow Hobby Horse Festival (May)r

Shetland Islands: Up-Helly-A' Celebration (January)

London:
Chelsea Flower Show (May),
Trooping the Colour (June),
Notting Hill Carnival (August),
State Opening of Parliament
(November or December),
Lord Mayor's Procession (November)

Liverpool: Grand National Steeplechase (April)

Wales: Royal National Eisteddfod (August)

Ladies Day at the Ascot race course is a major social event in June. Women traditionally wear extravagant clothing and hats on that day. Men wear morning coats and top hats.

The Last Night of the Proms celebrates British tradition with patriotic music in the Royal Albert Hall in London. The program takes place on the final evening of a series of summer concerts called promenades, or "proms."

EARLY HISTORY

After the Romans conquered England, they built walls to protect their settlements from the warlike peoples of Scotland. Hadrian's Wall, the most famous of these fortifications, was built by Emperor Hadrian in the A.D. 120's. It extended from Solway Firth to the mouth of the Tyne River.

In A.D. 43, Roman armies began their conquest of the Celtic tribes of Britannia, as the island of Great Britain was then called. The Romans conquered what is now England and Wales, but they never completely conquered Scotland and did not invade the small Celtic kingdoms of Ireland.

The Romans withdrew in the 400's. Invading Germanic tribes, especially the Angles, Saxons, and Jutes, then set up kingdoms throughout southern and eastern England. They eventually formed seven kingdoms, called the Heptarchy.

Egbert, king of Wessex from 802 to 839, was the last king to control the Heptarchy. He is often considered the first king of England. Farther north, Kenneth I MacAlpin, the king of the Scots, also became king of the Picts around 843. He established Alba, first united kingdom in Scotland.

In the late 700's, Danish Vikings began raiding the British Isles. From the 800's to early 1000's, they conquered large areas. In 886, Alfred the Great, king of Wessex, pushed the Viking settlers into the northeastern third of England.

In 1066, Harold, the last Saxon king, was killed at the Battle of Hastings by Normans led by William, duke of Normandy in France. William was then crowned king of England, and he became known as William the Conqueror. He rewarded his followers with lands in England. Starting in the late 1100's, some Anglo-Norman barons also began carving out territories in Ireland.

Power struggles

During the late 1000's and early 1100's, the English government was centralized, and its courts were reformed by several strong kings, like Henry II. In 1215, a group of barons and church leaders drew up the Magna Carta (Great Charter) and forced King John to agree to it. The document limited royal power and gave the nobles rights that later became important to all the people.

In the late 1200's, Parliament began to gain greater importance. In 1295, King Edward I called a meeting of town representatives, nobles, and church leaders. The meeting became known as the "Model Parliament" because it set the pattern for later Parliaments.

In 1282 and 1283, Edward I conquered Wales. He attempted to bring Scotland under English control, too, but the Scots rebelled repeatedly. The Scots' victory over Edward II in 1314 at Bannockburn assured Scotland's independence.

The 1300's and 1400's were years of conflict in England. The Hundred Years' War with France (1337-1453) was followed by the Wars of the Roses (1455-1485), a bitter struggle for control of the Crown. King Henry VII emerged as England's new ruler in 1485 and returned the country to stability.

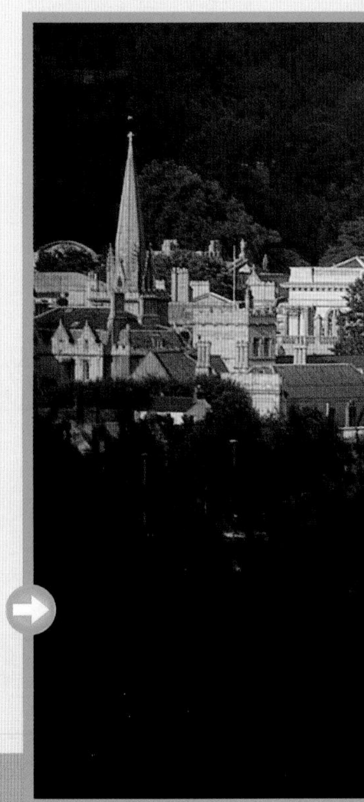

Oxford University developed during the 1100's. Cambridge University grew up in the 1200's. The two schools helped make England an important center of learning.

TIMELINE

A.D. 43	Romans invade Britannia.
400's	Germanic tribes invade Britain.
c. 843	Kenneth I MacAlpin establishes Alba, the first united kingdom in Scotland.
886	Alfred the Great, King of Wessex, defeats the Danes.
1066	Normans win the Battle of Hastings and conquer England.
1215	English barons force King John to sign the Magna Carta.
1282-1283	England conquers Wales.
1295	Edward I forms the Model Parliament.
1301	Edward I gives the title Prince of Wales to his son.
1314	Scotland keeps its independence by defeating Edward II at the Battle of Bannockburn.
1337-1453	England and France fight in the Hundred Years' War.
1455-1485	Two royal families fight for the throne in the Wars of the Roses.
1534	Henry VIII has Parliament make him the supreme head of the church in England.
1536	Henry VIII unites England and Wales.
1558-1603	Elizabeth I reigns during the Golden Age of English history.
1588	The English defeat the Spanish Armada.
1599-1608	Shakespeare writes *Hamlet, Julius Caesar, King Lear, Macbeth, and Othello.*
1603	England and Scotland are joined under one king, James I.
1649	Charles I is executed. England becomes a republic.
1653-1659	Oliver Cromwell and his son rule England.
1660	Parliament restores the monarchy.
1688	The Glorious Revolution ends the rule of James II.

Queen Elizabeth I (1533-1603)

William Shakespeare (1564-1616)

Oliver Cromwell (1599-1658)

Reformation to 1688

Henry VIII initiated the Protestant Reformation in England in 1529. The king had Parliament pass a law in 1534 naming the king, not the pope, supreme head of the church in England. This move led to the formation of the Church of England, independent of the Roman Catholic Church. In 1536 and 1543, Parliament passed Acts of Union that fully united the government systems of England and Wales.

Elizabeth I became queen in 1558, and her reign is called the Golden Age of English history. Explorers, such as Sir Francis Drake, and writers, such as William Shakespeare, brought glory to the Elizabethan era. In 1588, England defeated the Armada, a huge invasion fleet sent by Spain, the most powerful nation in Europe.

After Elizabeth died in 1603, her cousin, James VI of Scotland, also became James I of England. But each country kept its own laws and Parliament.

Several kings in the 1600's sought to exercise absolute power. Charles I did not allow Parliament to meet from 1629 to 1640. As a result, civil war broke out in 1642, and Charles was beheaded in 1649. England became a republic led by army commander Oliver Cromwell.

In 1660, the monarchy was restored under Charles II. Civil war once again threatened under James II, who became king in 1685. His efforts to restore Catholicism to England were defeated during the Glorious Revolution in 1688, when leading English politicians invited James's daughter Mary and her husband, William of Orange, ruler of the Netherlands, to rule England. William invaded England with Dutch forces, and James fled to France.

MODERN HISTORY

The Great Exhibition of 1851, held in London, reflected a national spirit of optimism and energy. At that time, the United Kingdom was the most powerful and technologically advanced nation in the world.

At the same time that Parliament turned over the throne of England to William and Mary in 1689, it moved to limit the power of the monarchy. The new rulers agreed to accept the Bill of Rights, which granted the people basic civil rights and curbed the king's power in such matters as taxation and keeping a standing army.

In 1707, under Queen Anne, the Act of Union joined England, Wales, and Scotland as a "united kingdom of Great Britain." By this time, Parliament had won a controlling influence over the monarchy. After Queen Anne died in 1714, her cousin, George, a German prince, became king. However, George I spoke little English, so his chief minister, Sir Robert Walpole, controlled the council of ministers. Thus began the British Cabinet system of government. Walpole is considered the country's first prime minister.

The British Empire

At the end of the Seven Years' War (known in America as the French and Indian War) in 1763, Great Britain acquired many of France's territories in North America and India. The American Revolution (1775-1783) cost Britain the most valuable part of its empire—the colonies that became the United States. However, shortly after the war ended, Britain grew richer than ever before through trade with the new nation. In the late 1700's, the Industrial Revolution began in Britain and eventually made that country the richest nation in the world.

In 1793, Britain once again went to war against France, which from 1799 was led by Napoleon I, also known as Napoleon Bonaparte. In 1815, the British victory by the Duke of Wellington in the Battle of Waterloo ended the Napoleonic Wars.

From 1837 until 1901, Queen Victoria enjoyed the longest reign in British history. The British Empire reached its height, and the nation became an industrial powerhouse.

The 1900's

The United Kingdom was on the winning side both in World War I (1914-1918) and World War II (1939-1945). But the wars devastated the country's armed forces and its economy.

In 1920, the United Kingdom divided Northern Ireland from the rest of Ireland to create separate governments in each. The south became self-governing in 1921 and a completely separate republic in 1949.

Between 1940 and 1980, about 40 British colonies became independent nations. Most remained associated with the United Kingdom through the Commonwealth of Nations. Meanwhile, ties with the rest of Europe grew closer in 1973, when the United Kingdom joined the European Community, now known as the European Union.

In 1997, Prime Minister Tony Blair called for referendums in which the voters of Scotland and Wales approved the formation of their own legislatures to handle many local issues. A 1998 peace agreement in Northern Ireland also provided for a local legislature and helped bring stability to that region.

TIMELINE

1689	William and Mary accept the Bill of Rights and become joint rulers of England.
1707	The Parliament of England and Wales and the Parliament of Scotland both pass the Act of Union, creating a "united kingdom of Great Britain."
1756-1763	Britain wins many of France's possessions in the Seven Years' War (also known as the French and Indian War).
1775-1783	Britain loses its American colonies in the American Revolution.
1793-1815	Britain defeats France in the Napoleonic Wars.
1801	Act of Union unites Great Britain and Ireland to form the United Kingdom of Great Britain and Ireland.
1837-1901	The British Empire reaches its height during the Victorian Age.
1914-1918	The United Kingdom and the Allies defeat Germany and the other Central Powers in World War I.
1921	The southern part of Ireland becomes a separate nation. The United Kingdom, including Northern Ireland, becomes the United Kingdom of Great Britain and Northern Ireland.
1931	Commonwealth of Nations is established.
1939-1945	The United Kingdom led by Prime Minister Winston Churchill and the Allies defeat Germany and the Axis Powers in World War II.
1947	India and Pakistan gain their independence.
1973	The United Kingdom becomes a member of the European Community, now known as the European Union.
1982	British troops defeat Argentine troops in battles for control of the Falkland Islands.
1991	British troops take part in the Persian Gulf War of 1991.
1994	Tunnel running under the English Channel opens, connecting the United Kingdom and France.
1997	Scotland and Wales vote to set up their own legislatures.
1998	Peace agreement in Northern Ireland includes plans for a legislative assembly for Northern Ireland.
2001	The United Kingdom joins the United States and other countries in a campaign against terrorism. British troops take part in military strikes in Afghanistan.
2003-2009	British troops are involved in the Iraq War.

Duke of Wellington (1769-1852)

Queen Victoria (1819-1901)

Sir Winston Churchill (1874-1965)

During the London Blitz in 1940, whole blocks of buildings were destroyed. Many people slept in air-raid shelters and subway stations during the bombings, which lasted for 57 straight days and nights.

During the early 2000's, the United Kingdom became part of an international campaign against terrorist activity. The British participated, alongside the United States and other nations, in military campaigns in Afghanistan and Iraq. The United Kingdom also sent peacekeeping forces to these and other places in an effort to increase security and stability in regions of unrest.

The first car rolls off a Channel Tunnel train on Dec. 22, 1994. Work on the tunnel, a railroad between England and France that runs under the English Channel, lasted for seven years.

THE UNITED STATES

The United States of America is a large and diverse nation that spans the middle of North America from the Atlantic Ocean to the Pacific Ocean. It also includes the huge state of Alaska in the northwest corner of North America and the island state of Hawaii, far out in the Pacific. The United States is the fourth largest country in the world. Only Russia, Canada, and China are larger. Only China and India have more people.

America's landscape is as varied as it is vast. It ranges from the warm, sunny beaches of California and Hawaii to the frozen northlands of Alaska, from the flat Midwestern plains to the towering Rocky Mountains, and from the swamps of Florida's Everglades to the desert of California's Death Valley.

The United States also has rich and varied natural resources. Some of the most fertile soil on Earth covers much of the Midwest. Vast forests flourish in the Northeast, Southeast, and Northwest. Under the ground lie huge deposits of valuable minerals. Excellent water transportation routes are provided by the five Great Lakes and by many rivers, including the Mississippi River system, which flows through the heart of the country. These resources have helped make the United States one of the world's most highly developed and productive nations. The people of the United States enjoy one of the highest standards of living in the world.

Most Americans are descendants of European settlers who arrived in North America after the 1500's. Until then, the area that is now the United States was largely a wilderness. Small groups of Native Americans lived on the land between the Atlantic and the Pacific. In addition, Inuit (also called Eskimos) lived in what is now Alaska, and Polynesians lived in Hawaii.

For many Europeans, this New World offered an opportunity to build a new and better life. Spanish settlers came to Florida and the Southwest. English people and other Europeans settled along the East Coast. In 1776, the British colonies in North America declared their independence from England. They founded a nation based on freedom and economic opportunity.

Through the years, immigrants from around the world have come to the United States. Except for black Africans who were brought in as slaves, these immigrants came seeking the rights and the opportunities that had become part of the American way of life. Since 1886, the Statue of Liberty in New York Harbor has welcomed many of these people to the New World.

Today, the United States is sometimes called a "nation of immigrants." Some groups have suffered socially and economically, but through the laws passed by their elected representatives, the people of the United States continue to seek "liberty and justice for all."

THE UNITED STATES TODAY

The Constitution of the United States, written more than 200 years ago, established the United States as a federal republic. Federal means that power is shared between a national government and state governments. In a republic, the people vote for leaders to represent them and guard their rights.

The federal government

The writers of the U.S. Constitution separated the federal government's power among three branches—the executive, the legislative, and the judicial.

The executive branch is headed by the president and includes executive departments and independent agencies. As chief executive, the president enforces federal laws, commands the armed forces, and conducts foreign affairs. The people elect the president to a four-year term through the Electoral College. No president can serve more than two terms.

The legislative branch is represented by the Congress, which is made up of two houses—the House of Representatives and the Senate. It also includes various agencies. The legislative branch makes the laws of the nation. Its powers include raising money through taxes, regulating trade between states, and declaring war.

The number of representatives each state sends to the House of Representatives depends on the population of the state. However, regardless of its size or population, each state sends just two senators to Congress.

The judicial branch of the United States is a system of courts. The highest court in the land is the Supreme Court of the United States. The nine justices of the Supreme Court interpret the Constitution and hear cases that involve federal laws.

In addition to these specific powers, each branch of government has powers that check or balance the powers of the other two branches. For example, the president can veto bills from Congress and appoint the justices of the Supreme Court. Congress can override presidential vetoes and organize the federal courts. And

FACTS

Official name:	United States of America
Capital:	Washington, D.C.
Terrain:	Vast central plain, mountains in west, hills and low mountains in east; rugged mountains and broad river valleys in Alaska; rugged, volcanic topography in Hawaii
Area:	3,616,240 mi² (9,366,014 km²)
Climate:	Mostly temperate, but tropical in Hawaii and Florida, arctic in Alaska, semiarid in the great plains west of the Mississippi River, and arid in the southwest; low winter temperatures in the northwest occasionally lessened by warm winds from the eastern slopes of the Rocky Mountains
Main rivers:	Mississippi, Ohio, Missouri, Colorado
Highest elevation:	Mount McKinley in Alaska, 20,320 ft (6,194 m)
Lowest elevation:	Death Valley in California, 282 ft (86 m) below sea level
Form of government:	Republic
Head of state:	President
Head of government:	President
Administrative areas:	50 states, 1 district
Legislature:	Congress consisting of Senate with 100 members serving six-year terms and House of Representatives with 435 members serving two-year terms
Court system:	Supreme Court
Armed forces:	1,539,600 troops
National holiday:	Independence Day - July 4 (1776)
Estimated 2010 population:	310,299,000
Population density:	86 persons per mi² (33 per km²)
Population distribution:	81% urban, 19% rural
Life expectancy in years:	Male, 75; female, 81
Doctors per 1,000 people:	2.6
Birth rate per 1,000:	14
Death rate per 1,000:	8
Infant mortality:	7 deaths per 1,000 live births
Age structure:	0-14: 20%; 15-64: 67%; 65 and over: 13%
Internet users per 100 people:	71
Internet code:	.us
Languages spoken:	English, Spanish
Religions:	Protestant 51.3%, Roman Catholic 23.9%, Mormon 1.7%, Jewish 1.7%, other 21.4%
Currency:	United States dollar
Gross domestic product (GDP) in 2008:	$14.266 trillion U.S.
Real annual growth rate (2008):	1.3%
GDP per capita (2008):	$46,798 U.S.
Goods exported:	Chemicals, electronics, food, machinery, plastic products, transportation equipment
Goods imported:	Chemical products, crude oil, electronics, iron and steel, machinery, transportation equipment
Trading partners:	Canada, China, Germany, Japan, Mexico

the Supreme Court can declare executive orders and laws unconstitutional. This system of checks and balances prevents any one branch from becoming too powerful.

The White House is the official residence of the president of the United States. The president not only lives in the White House but also works there.

The state governments

Each of the 50 states can exercise powers given to the states—or not denied them—by the Constitution. Each state has its own constitution, with an executive branch (headed by a governor), a legislative branch (headed by a state legislature), and a judicial branch (headed by a state supreme court).

Some of the powers exercised by the states include the maintenance of law and order, administration of health and welfare services, and most regulation of business. The states also have the major responsibility for public education.

Government spending

Government plays a major role in the U.S. economy. Federal, state, and local governments employ about 15 percent of all workers and buy a third of all the goods and services produced in the United States. The federal government spends part of its budget on social security benefits for American people, and both federal and state governments provide medical and welfare aid for the elderly and for the needy.

A statue of Abraham Lincoln is part of the Lincoln Memorial in Washington, D.C. The beautiful white marble monument stands at the end of the National Mall and honors the democratic ideals of the 16th president of the United States. From a humble birth, Lincoln rose to the presidency, led the nation during the Civil War (1861–1865), and freed the slaves.

THE FIFTY STATES

The 50 states that together form the United States cover an area of more than 3-1/2 million square miles (9 million square kilometers). The northernmost point is Point Barrow, Alaska, and the southernmost point is Ka Lae, Hawaii. The easternmost point lies at West Quoddy Head, Maine, and the westernmost point at Cape Wrangell on Attu Island in Alaska.

The 48 states that border one another between the Atlantic and Pacific oceans stretch across four time zones and 2,807 miles (4,517 kilometers) at the greatest distance. They differ greatly in size, ranging from huge Texas to tiny Rhode Island. Alaska, which is geographically separate from the other states, has the largest area of all the states.

The 50 states are political divisions of the United States. Because the United States has a federal system of government, the states hold many of the powers assumed by national governments in most other countries. For example, states have a great deal of control over education, and they can pass many civil and criminal laws. Some states can be compared with many small countries in size, population, and economic output, as well as government.

In addition to the 50 states, the District of Columbia, the seat of the federal government, is an important part of the United States. The District of Columbia lies along the Potomac River between Virginia and Maryland. The 68-square-mile (177-square-kilometer) area was set aside by the government for the U.S. capital city of Washington. The city is best known by the name Washington, D.C. The United States also has island possessions and territories in the Caribbean Sea and Pacific Ocean.

The states of the United States are often divided into regions that share geography, climate, economy, traditions, and history with one another. Because America was largely settled from east to west, states on the East Coast and along the southern shores of the Great Lakes have long been centers of dense population. But in more recent times, people have been attracted by the mild climate and growing business opportunities in the Southern and Western Sun Belt states. Today, California has more residents than any other state.

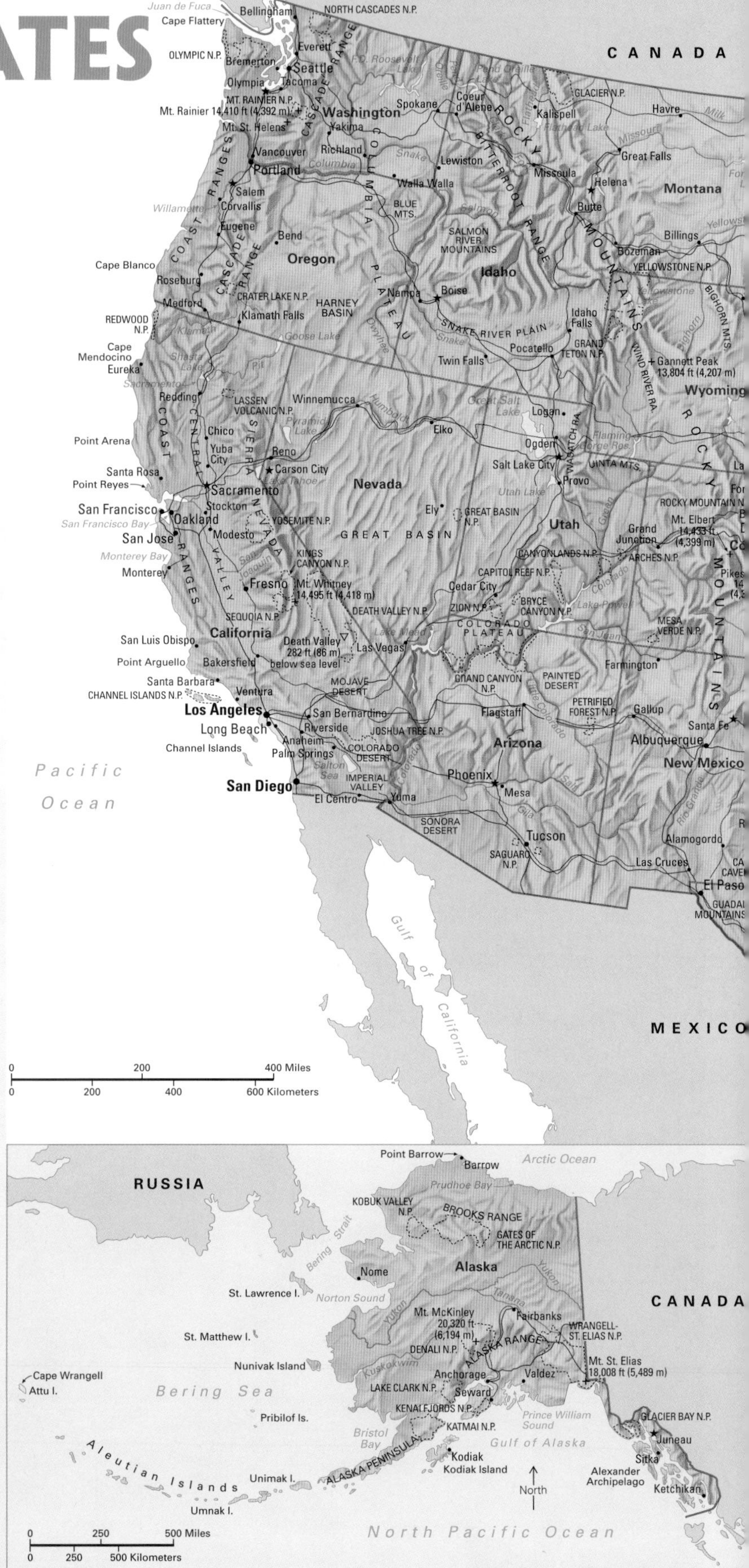

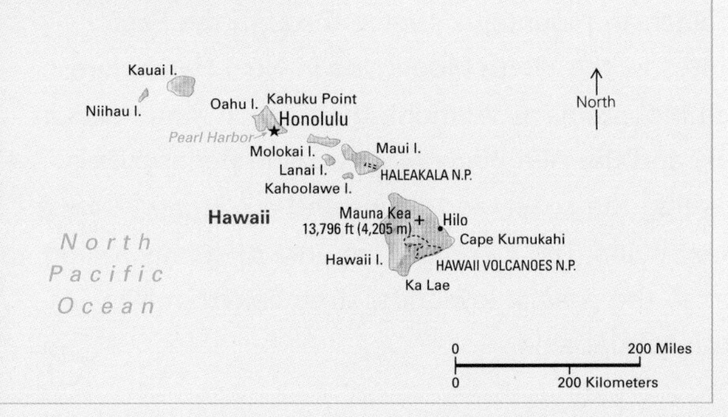

Fifty states make up the United States of America. The far Northeastern states, called New England, and the Middle Atlantic States were the first to be settled by Europeans. Today, great cities like Boston and New York are located there. West of the Appalachians, the states of the Midwest spread out over the plains of the Mississippi River system. This area is rich farm country, but it also became heavily industrialized, starting in the late 1800's. The warm South, once primarily an agricultural region, now has growing businesses and industries. Moving west, the Great Plains slope upward to the Rocky Mountains and the dry plateaus of the Southwest. This area offers some of the most spectacular scenery in the country. The Pacific Coast States have dense forests, rugged mountains, vast deserts, and wild ocean shores. The mild climate attracts tourists and new residents. Alaska and Hawaii, geographically separate from the other 48 states, both became states in 1959.

EAST COAST STATES

The states that occupy what is now the northeast corner of the country were part of the original Thirteen Colonies. Along with the Southern Colonies, they declared their independence on July 4, 1776, and the United States was born.

These East Coast States are often considered two smaller regions: New England, consisting of Connecticut, Maine, Massachusetts, New Hampshire, Rhode Island, and Vermont; and the Middle Atlantic States, consisting of New Jersey, New York, and Pennsylvania.

Many New Englanders are descendants of the English Puritans who settled the Massachusetts Bay region in the 1600's. Dutch immigrants originally settled in what is now New York. Their city of New Amsterdam became the budding city of New York when England took control of the colony. William Penn, a Quaker, founded Pennsylvania as a place where people of every faith could enjoy religious freedom.

The U.S. Capitol in Washington, D.C., attracts visitors from all over the world. The nation's capital is at the southern end of a huge megalopolis that includes Boston and New York City.

New Hampshire
Vermont
Maine
Massachusetts
New York
Rhode Island
Connecticut
Pennsylvania
New Jersey

The Old State House is a historic landmark in Boston. Built in 1713, the State House was the seat of the colonial government. Today, the towers of Boston's downtown area rise around it.

Varied land and a varied economy

The Appalachian Mountains stretch through the East Coast States as the White Mountains in New Hampshire, the Green Mountains in Vermont, the Catskill Mountains in New York, and the Allegheny Mountains in Pennsylvania. Lowlands hug the coast, and plains and a plateau lie west of the mountains. The Piedmont, an area of gently rolling hills west of the coastal lowlands, stretches from New York south to Alabama.

New England was the nation's first industrial center, and manufacturing is still important in the region. Much of New England is too hilly or rocky to grow crops on a

large-scale, commercial basis, but the region has many productive dairy and poultry farms, fruit orchards, and greenhouses. It is also famous for its maple syrup. New England's land and history contribute to the economy in still another way. Many tourists visit the region's historic sites, picturesque rural villages, fishing harbors, and fine ski resorts.

The deepwater harbors of the Middle Atlantic States—and the states' large populations—help make this area a major center of trade. Factories produce a variety of products, and coal is mined in Pennsylvania. Farms dot the hillsides and plains in these states. Tourists are drawn to this area too, attracted by its forested mountains, scenic lakes and rivers, big cities, and historic sites such as Gettysburg, the Pennsylvania community that is famous as the site of the American Civil War's Battle of Gettysburg.

Another attraction for tourists is the Appalachian National Scenic Trail, a hiking path almost 2,000 miles (3,200 kilometers) long. The trail begins in Maine and stretches through New England and the Middle Atlantic States into the Deep South.

The cities

The East Coast States have developed the largest *megalopolis* (great city) in the United States. A megalopolis is a region where two or more metropolitan areas have grown together. A huge megalopolis includes Boston, New York City, Philadelphia, Baltimore, and Washington, D.C.

Boston, with more than half a million people, is one of the oldest and most historic U.S. cities. It is New England's leading business center as well as a national center of learning.

New York City, with about 8 million people, is the largest U.S. city. It is the center of a metropolitan area of nearly 20 million residents. As a center of international business and culture, New York affects much of what happens around the world.

Philadelphia was the birthplace of the United States. The Declaration of Independence and the Constitution of the United States both were signed in Philadelphia's historic Independence Hall. Today, the city has about 1 ½ million people.

Scarecrows watch over pumpkins for sale at a farm in Vermont. Famous for its maple syrup, its scenic autumn beauty, and its skiing, Vermont is one of the least populated of the 50 states.

Rockefeller Center is a famous building complex in the Manhattan borough of New York City. The center consists of 19 commercial buildings dominated by the 70-story GE Building, originally known as the RCA Building.

MIDWESTERN STATES

Twelve states that lie in the heart, or center, of the United States are sometimes called "the heartland." The Midwestern States of Illinois, Indiana, Iowa, Kansas, Michigan, Minnesota, Missouri, Nebraska, North Dakota, Ohio, South Dakota, and Wisconsin cover a vast area of fairly flat and very fertile land. They produce food for the country and for the rest of the world.

Wisconsin dairy farms, with their well-kept outbuildings and green fields, make up a typical Midwest scene. Wisconsin's huge output of milk, cheese, and butter has earned this Midwest state the nickname America's Dairyland.

The eastern states of this region—Illinois, Indiana, Michigan, Minnesota, Ohio, and Wisconsin—lie around four of the five Great Lakes. These lakes were the chief route taken by explorers and settlers traveling into what is now the Midwest. Thousands of other pioneers from the Eastern United States sailed down the Ohio River on large flatboats to Midwestern settlements.

The Great Plains—a vast grassland—stretches across the western part of the Midwest. American Indians who lived there fought to keep whites from settling on their lands, but they were defeated by U.S. forces. The U.S. government forced the Indians to sign treaties that cleared the way for whites to live there.

Farmers in Nebraska celebrate their corn harvest. Once called the "Great American Desert," Nebraska is now a leading farm state. Its nickname is the Cornhusker State.

Ranchers moved onto the land first, and soon the Great Plains was a vast cattle empire. Railroads expanded westward to provide transportation to eastern markets. Farmers known as *homesteaders* followed the ranchers onto the Great Plains. Such new inventions as barbed wire and improved windmills, along with free land, helped the homesteaders turn the plains into a productive agricultural region, where both cattle and crops could be raised.

The breadbasket

The Midwest is noted for its vast expanses of fertile soil. The Great Plains represents one of the world's chief wheat-growing areas. In addition to wheat, farms in the Midwest also produce huge quantities of alfalfa, barley, corn, oats, and rye, as well as dairy products and livestock. Iowa and Illinois are the nation's leading soybean-producing states.

Downtown Chicago extends about 25 blocks along the west shore of Lake Michigan. The Chicago River winds through the downtown area.

Chemical carrier barges pass through a lock on the Mississippi River at Alton, Illinois. The Mississippi River system, along with the Great Lakes and many railroads, gives the Midwest an excellent transportation network.

Many of the people who live in the rural areas of the Midwest today are the descendants of German, British, Swedish, and Norwegian settlers.

Industry, services, and cities

Although the Midwest is a major farming region, it also has large industrial cities and an important service economy. The Great Lakes were important in the industrial development of the United States and the Midwest, especially for the steel industry. These deep lakes provide a fast water route for ships carrying iron ore from ports in northern Michigan, Minnesota, and Wisconsin to steel mills in Indiana and Ohio. The Great Lakes are also the best way to ship the huge wheat crops from the Great Plains.

The U.S. automobile industry has been centered in Michigan since the early 1900's. Detroit, called the *Motor City,* was the birthplace of the first Ford car.

The railroads that helped settle the plains turned Chicago into an important rail hub. The city now ranks as one of the nation's leading transportation centers. Chicago is the only place where the Great Lakes connect with the Mississippi River system. This energetic city also has some of the world's tallest buildings and one of the busiest airports.

Other large industrial cities located along the Great Lakes include Milwaukee and Cleveland. Major Midwestern cities away from the Great Lakes include Minneapolis, St. Louis, Kansas City, and Indianapolis. Many descendants of immigrants from northern, eastern, and southern Europe, as well as African Americans and Hispanic Americans, live in the big cities of the Midwest.

SOUTHERN STATES

The South is made up of the states of Alabama, Arkansas, Delaware, Florida, Georgia, Kentucky, Louisiana, Maryland, Mississippi, North Carolina, South Carolina, Tennessee, Virginia, and West Virginia.

The Mason-Dixon Line—the east-west boundary line separating Pennsylvania from Maryland and part of West Virginia, and the north-south boundary line between Maryland and Delaware—came to be considered the dividing line between the North and South. Before the American Civil War (1861-1865), the southern border of Pennsylvania was also the boundary between the antislavery states to the north and the proslavery states to the south.

From the 1600's to the 1800's, cotton and tobacco, grown on large plantations, led to an economy in the South that depended on the labor of African slaves. In the North, however, the economy was based on trade and manufacturing. As a result, two very different cultures and lifestyles developed.

The North's opposition to slavery led to tension and, eventually, war. From 1860 to 1861, 11 Southern States broke away from the Union and formed a Confederacy. The bitter and bloody Civil War followed, ending with the South's defeat. About 260,000 of its men had been killed, and its economy was nearly destroyed. For decades, the South remained a fairly poor agricultural region. Not until the mid-1900's did its economy begin to change and grow stronger.

Today, many Southerners retain a strong sense of regional loyalty and take pride in the South's history and traditions.

New Orleans, with its quaint streetcars, historic French Quarter, and annual Mardi Gras festival, attracts millions of tourists each year. The city is also a major business, cultural, and shipping center of the South.

The land and people

The South is an area of rolling hills, mountains, and plains bordered by beaches along the Atlantic Ocean and the Gulf of Mexico. The Piedmont, which begins in New York, widens in Virginia and extends south into Alabama. Tobacco is widely grown in the Piedmont. The division between this gently rolling land and the Coastal Plain is the Fall Line, where rivers flowing from the west drop from higher, rocky ground to the sandy plain and then flow to the sea. Large cities have grown up along the Fall Line. They include Baltimore, Maryland; Columbia, South Carolina; Richmond, Virginia; Washington, D.C.; and Wilmington, Delaware.

The Appalachians extend from the North into the South as the Blue Ridge Mountains and the Great Smoky Mountains. Mining is important in these mountain areas, especially coal mining in Kentucky and West Virginia. Lumber from this re-

The lazy lower stretches of the Mississippi River break off into many small side channels as the river flows through bayous south of New Orleans. "Old Man River" is a major waterway for the South, linking the heartland of the United States with the rest of the world.

gion is shipped to furniture factories in North Carolina. Many of the people in the Appalachians, as elsewhere in the South, are descended from early English, Irish, and Scottish immigrants.

Flat or gently rolling plains areas along the coasts and in parts of Kentucky and Tennessee are forested or farmed. At one time, cotton was "king" in this Deep South land. Plantations spread over many acres, and slaves were numerous. Today, the descendants of those slaves form the largest minority group in the South.

The new South

The industrial boom that began in the mid-1900's greatly increased manufacturing in the South and improved its economy.

The moist subtropical climate of the South brings thousands of tourists to the region each year, especially in winter, when the North is cold and snowy. In recent years, many new residents have come to live and work in this Sun Belt region, and many retired people move to Florida.

Nighttime traffic flows on Ocean Drive in the South Beach area of Miami Beach. South Beach is one of the liveliest entertainment districts in Florida.

Atlanta is a center of trade and transportion for the southeastern United States. Atlanta is the capital and largest city of Georgia.

WESTERN INTERIOR STATES

The states in the Western interior contain some of the most spectacular scenery and natural wonders in the world.

The Western interior is made up of two regions: the Southwestern States of Arizona, New Mexico, Oklahoma, and Texas; and the Rocky Mountain States of Colorado, Idaho, Montana, Nevada, Utah, and Wyoming.

The beautiful Cliff Palace in southwestern Colorado was built by Pueblos, probably in the 1200's. Now part of Mesa Verde National Park, the dwellings help preserve the Native American heritage of the Southwest.

From the 1500's until the early 1800's, most of the Southwest was Spanish territory. The United States *annexed* (added) Texas in 1845, 10 years after Texans had staged a revolt against Mexican rule. The country paid Mexico for the rest of its southwestern land in 1848 and 1853. The region's history is reflected in the Hispanic influences seen throughout the Southwest.

Both the Southwestern and the Rocky Mountain States were part of the Wild West. Although the West was not really as wild as legend and movies suggest, cowboys, miners, stagecoach drivers, sheriffs, and homesteaders all played a part in this region's history.

The Southwest

The Southwestern States spread out over a vast area that is sometimes called the "wide-open spaces." There, cattle graze on huge ranches. In some parts of the Southwest, vast fields of cotton and other crops soak up the abundant sunshine. In other areas, barren desert stretches as far as the eye can see.

In the 1900's, refineries and factories making chemicals from petroleum products helped industrialize the Southwest. Petroleum has brought the region much of its wealth. Texas and Oklahoma are among the nation's leading producers of petroleum and natural gas.

Las Vegas Boulevard, known as the Strip, is lined with extravagant hotels and casinos, making the city one of the world's most popular tourist destinations.

Industrialization has led to a growing population, especially in urban areas where service industries are concentrated. Many of the fastest-growing U.S. cities are in the Southwest.

Many retired people move to the Southwest for its warm climate, while tourists come to enjoy the region's stunning natural beauty. The Grand Canyon draws millions of visitors each year. Arizona's Painted Desert, New Mexico's Carlsbad Caverns, and Big Bend National Park in Texas are just a few of the area's tourist attractions.

Many people come to visit ancient Native American ruins and observe today's Native American culture. The Southwest is home to Apaches, Navajos, and Hopis and other Pueblos. Native American art, pottery, jewelry, and blankets are highly prized.

Sixth Street in Austin, Texas, is the heart of the city's lively entertainment scene. The street is lined with historic houses and commercial buildings dating from the late 1800's and early 1900's. They house popular clubs, art galleries, cafes, restaurants, and hotels.

The Rocky Mountain States

The states of the Rocky Mountain region are named for the rugged, majestic range that cuts through the area. But the scenic landscape also includes deserts, plateaus, and plains.

Rich deposits of gold, silver, and other metals first attracted settlers to the Rocky Mountain region. Mining remains important, though manufacturing is now the chief source of income. Cattle and other livestock graze on dry, grassy ranges. Farmers grow a variety of crops, such as wheat, hay, and potatoes. Many people are now employed in service industries, such as real estate, health care, and wholesale and retail trade.

As in the Southwest, tourism is important to the area. Some visitors come to ski in Idaho's Sun Valley and Colorado's Vail and Aspen. Others come to camp, fish, hike, and enjoy the mountain and desert scenery.

Snow-dusted Boulder Valley in western Montana is Rocky Mountain country. Spruce, pine, and fir forests carpet the lower mountain slopes, but no trees grow above the timber line.

Valuable water

Water is scarce in this generally dry part of the country. Part of Utah and almost all of Nevada, for example, are in the Basin and Range country, one of the driest areas in the nation. The Colorado River flows across 1,360 miles (2,189 kilometers) of the country and carves the Grand Canyon. A number of dams, including Davis, Glen Canyon, Hoover, and Parker dams, stand along the river. They help prevent floods and erosion and provide electric power and water for much of the Southwest. So much water is stored in reservoirs and drained off by aqueducts and canals that the Colorado no longer reaches its mouth at the Gulf of California.

PACIFIC COAST STATES

The states that lie along the Pacific Coast— California, Oregon, and Washington—are known for their dense forests, rugged mountains, and dramatic ocean shore. Their beauty and relatively mild climate—cooler and wetter in the north, warmer and drier in the south—encourage an outdoor lifestyle enjoyed the year around by both residents and visitors.

Washington and Oregon were part of the Oregon Country, territory given to the United States by the United Kingdom in 1846. Settlers from the Midwest followed the Oregon Trail to this Pacific Northwest region.

California belonged to Spain and then to Mexico before the United States paid for it in 1848. The discovery of gold that same year brought thousands to the area in the Gold Rush. By 1850, California had enough people to be admitted to the Union as a state.

Land and people

Fertile valleys between the two parallel chains of Pacific Coast Ranges produce a large part of the nation's fruits, nuts, vegetables, and wine grapes. The Central Valley and Imperial Valley of California are especially productive farming areas.

The region also has abundant timber, minerals, and fish. Manufacturing activities in the region include the building of transportation equipment in Washington, the processing of wood in Oregon, and the production of electronic and computer equipment in California. Service industries, such as finance, health care, education, and retail trade, are centered in the region's urban areas. The region's large cities include Los Angeles, San Diego, San Jose, San Francisco, Portland, and Seattle. Hollywood, in California, is known worldwide for its movies and television shows.

Olympic National Park in Washington features some of the country's most scenic areas of unspoiled nature. The state also has two other national parks, Mount Rainier and North Cascades.

The fog-shrouded Golden Gate Bridge spans the entrance to San Francisco Bay. San Francisco is known for its picturesque cable cars, its bustling Chinatown, and also its earthquakes.

Waterfalls spill into the Oheo pools within the Haleakala National Park on the island of Maui in Hawaii. The park is also noted for an extinct volcano with colorful rock formations in its crater.

A stream of settlers to the Pacific Coast States began in the mid-1800's, and new residents have continued to pour in ever since. The California coast—from San Francisco through Los Angeles to San Diego—is a developing megalopolis.

Today, the region's population includes people of European, African American, and Mexican American ancestry. In fact, more people of Mexican origin live in the Los Angeles area than in any other U.S. city. More people of Asian ancestry make their home in the Pacific Coast States than in any other part of the United States. A large number of Native Americans live there as well.

Hawaii

Hawaii is the 50th state, the last state to be admitted to the Union. It can be considered a Pacific Coast State because its entire border is on the ocean. Hawaii lies in the middle of the North Pacific Ocean, about 2,400 miles (3,860 kilometers) from the U.S. mainland. The state is a string of 8 main islands and 124 smaller ones. About 75 percent of the state's people live on the island of Oahu, home to the city of Honolulu.

Hawaii is world famous for its beauty and pleasant climate. The original settlers were Polynesians, but many other national and ethnic groups have also contributed to Hawaii's colorful way of life. Not surprisingly, tourism is a major Hawaiian industry.

Giant sequoias reach for the sky in Sequoia National Park in California. The park's General Sherman Tree is one of the world's largest trees and one of the oldest living things on Earth.

ALASKA

The name *Alaska* comes from a word used by native people meaning *great land* or *mainland*. Alaska is by far the largest state in the United States. Spreading across the northwest corner of North America, it is separated from Washington state by about 500 miles (800 kilometers) of Canadian territory. Alaskans often refer to the rest of the continental United States as the "lower 48."

The Pacific Coast Ranges extend from the lower 48 through Canada and into Alaska, where they become the Alaska Range. The Rocky Mountains reach into Alaska as the Brooks Range. Alaska has almost all the active volcanoes in the United States, as well as the 16 highest peaks in the country. Mount McKinley rises 20,320 feet (6,194 meters) above sea level, the highest point on the North American continent.

The Alaskan wilderness attracts many tourists who love the outdoors. Huge forests cover about a third of the state. The middle section has low, rolling hills and broad, swampy river valleys. The southern coast is cut by hundreds of small bays and narrow, steep-sided inlets called *fiords*, while thousands of glaciers fill Alaska's mountain valleys and canyons.

The far northern treeless plain is covered with grasses and flowers in summer, but permanently frozen ground called *permafrost* lies underneath. This plain, called the *tundra*, is the summer grazing grounds of huge herds of caribou.

Other wild animals include polar bears, grizzly bears, deer, elk, moose, and mountain goats. The world's largest herd of northern fur seals live on the Pribilof Islands in summer. The waters off Alaska's shores are rich in salmon, halibut, and crab. Alaskan fishermen catch huge quantities of fish each year.

Many people think of Alaska as very cold and snowy, but the state has a great variety of climates. Warm ocean winds give southern Alaska a fairly mild climate. Parts of the southeast,

The Orthodox Church on Kodiak Island is a reminder of the time when Alaska was a Russian colony.

The Iditarod is the world's most famous sled dog race. The annual race starts on the first Saturday of March in Anchorage and ends in Nome. It crosses the Alaska and Kuskokwim mountain ranges, heading northwest across the state and then north along the Bering Sea coast to Nome.

Alaska is the largest state of the United States in area. Alaska's general coastline is 6,640 miles (10,686 kilometers) long. Most of it is along the Pacific Ocean but some is along the Arctic Ocean. Almost a third of Alaska lies above the Arctic Circle. The Alaskan mainland's most western point is only 51 miles (82 kilometers) from Russia. No other part of North America is closer to Asia.

called the Panhandle, get heavy precipitation. Inland, Alaska has cold winters and cool summers and is fairly dry. The Alaskan Arctic has even colder winters, cooler summers, and less precipitation.

The climate and soil as far north as the Arctic Circle allow farmers to raise livestock and grow barley, potatoes, and other hardy crops. Crops grow rapidly because the summer sun shines 20 hours a day. At Point Barrow in the Arctic, from May 10 to August 2, the sun never sets.

Alaska's people

When Europeans first arrived in Alaska in the 1700's, three groups of native people were living there. Inuit lived near the coast in the far west and north. Aleuts—closely related to the Inuit—lived on the Aleutian Islands and the Alaska Peninsula. American Indians lived on the southeastern islands, along the coast, and in the interior.

A Russian expedition in 1741 landed in Alaska and left with sea otter furs. Russian fur traders soon followed, and the area became Russian territory.

In 1867, however, Russia sold Alaska to the United States for $7.2 million. Some Americans thought the purchase was foolish. They called Alaska *Seward's Folly* or *Seward's Icebox*, after William Seward, the U.S. official who agreed to buy it.

Today, Alaska has a population of more than 600,000. Nearly two-fifths of Alaska's people were born in Alaska. Many of those born in other states are members of the United States armed forces who have been assigned to Alaska. Most of Alaska's people live in Anchorage (the largest city), Fairbanks, or cities on the southeast coast. Alaska is the most thinly populated state in the country.

The oil boom

One of the greatest oil discoveries of all time was made in 1968 at Prudhoe Bay on the Arctic Coast. In 1977, a pipeline was completed to carry the oil about 800 miles (1,300 kilometers) from Prudhoe Bay on the Arctic Ocean to Valdez on the southern coast, and oil production began.

Oil has boosted the state's economy and created many jobs, but oil production can cause problems too. In 1989, a tanker, the *Exxon Valdez*, struck a reef and caused a major oil spill in Prince William Sound. The oil killed wildlife and polluted beaches and fishing grounds.

A tour boat eases past Muir Glacier in Glacier Bay National Park. Many tourists take such cruises along Alaska's southeast coast.

NATIONAL PARKS

In 1807, an American trapper came across a wonderland in the northwest corner of what is now Wyoming. He and later trappers returned with stories of spouting geysers, deep gorges, thundering waterfalls, eerie hot springs, mineral deposits, and bubbling pools of mud.

In 1870, an expedition went into the area to see if the fantastic stories were true. By 1872, Congress had created a park out of the land. The park was named Yellowstone, for the river that runs through it and for the yellow rocks that lie along the river.

Yellowstone became the world's first national park.

Natural wonders and monuments

The United States is rich in natural wonders like Yellowstone, as well as famous historic places and recreation sites. Since 1872, the government has set aside about 400 of these areas as national parklands. The parklands include parks, monuments, historic sites, memorials, cemeteries, seashores, lakeshores, scenic riverways, and battlefields.

Yosemite National Park, in east-central California, was one of the first national parks created in the United States. The park is known for its spectacular scenery. Yosemite Falls is North America's highest waterfall.

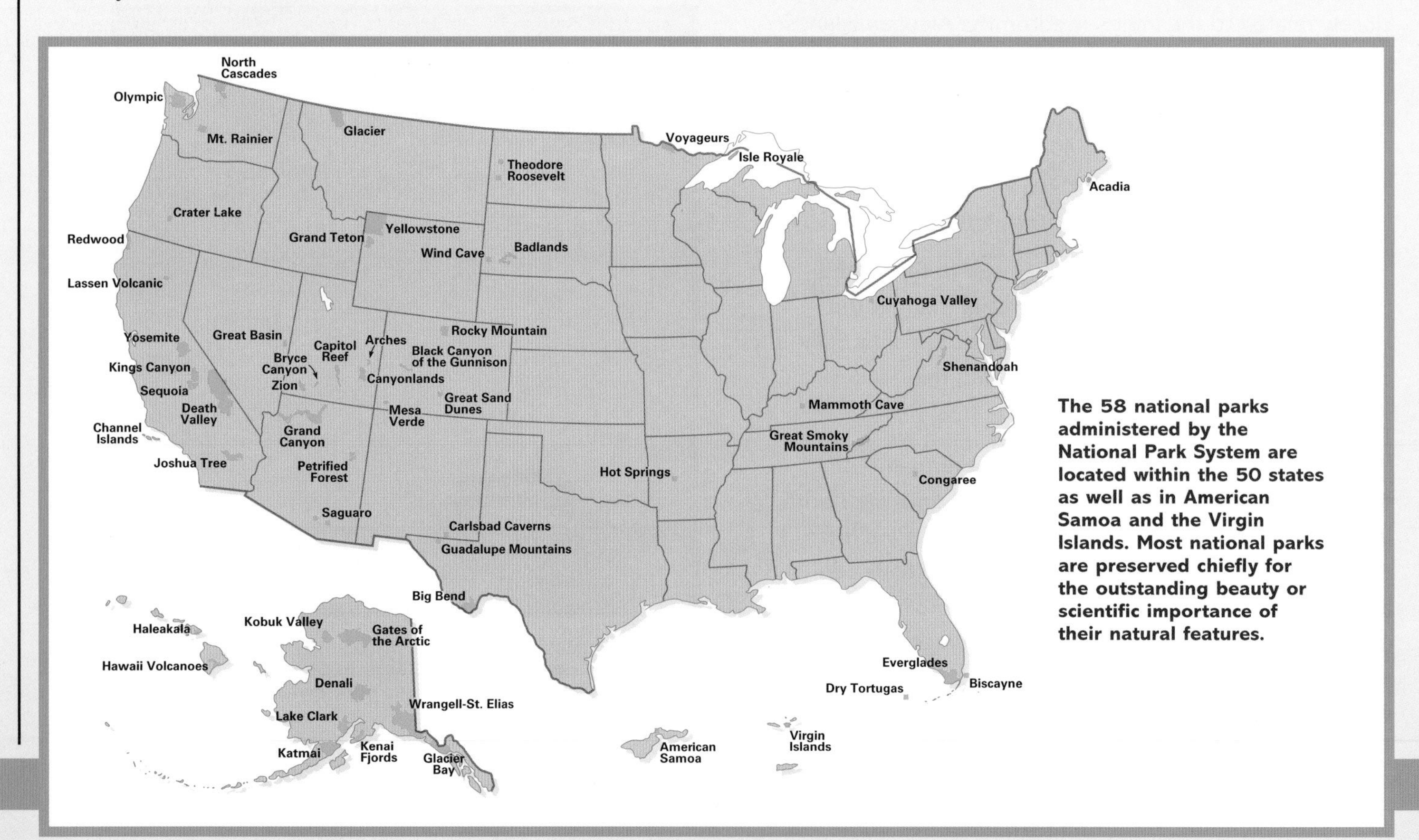

The 58 national parks administered by the National Park System are located within the 50 states as well as in American Samoa and the Virgin Islands. Most national parks are preserved chiefly for the outstanding beauty or scientific importance of their natural features.

Most national parks have been set aside for their beauty or scientific importance. Acadia, in Maine, features a wild rocky coastline. Shenandoah, in Virginia, has the hardwood-forested Blue Ridge Mountains. In Grand Teton, south of Yellowstone, are rugged mountains and herds of elk. Crater Lake, in Oregon, features a deep-blue lake in a dead volcano and colorful lava walls almost 2,000 feet (610 meters) high.

Historic national monuments include ancient ruins, such as the towns and mounds built by ancient Native Americans at Ocmulgee in Georgia, as well as battlefields like Gettysburg in Pennsylvania. Abraham Lincoln's birthplace is a historic site, and so is the White House.

National recreational areas range from Lake Mead on the Arizona-Nevada border to Sleeping Bear Dunes in Michigan. These areas provide land and water resources for such outdoor activities as fishing, horseback riding, sailing, and water-skiing.

The National Park Service

Almost all the national parklands are managed by the National Park Service, a bureau of the U.S. Department of the Interior. Park rangers patrol the lands to protect them from damage and also provide services for tourists.

To preserve the parks, the balance of nature must be maintained, so that plant and animal life is left as undisturbed as possible. Fishing is allowed, but hunting, lumbering, and mining are banned in most areas. The Park Service also lets many natural fires run their course.

The National Park Service encourages recreational activities in the parklands as long as the activities do not disturb the surroundings. Rangers try to teach people about the plant and animal life and the natural processes that shaped the land. In historic parklands, staff members sometimes restore old buildings and wear costumes that reflect America's colorful past.

The large numbers of visitors have caused problems in some parklands. Overcrowding can result in heavy car traffic, air pollution, dirty streams, and jammed campgrounds. Careful management is needed to meet visitors' demands without damaging the land.

Castle Geyser is one of more than 300 geysers in Yellowstone National Park. The park is one of America's largest wildlife preserves. Bears, elk, and bison roam freely through Yellowstone's forests and meadows.

In Carlsbad Caverns, rock formations called *stalactites* hang from the cave roofs, while *stalagmites* rise from the ground. Some passages in the caverns are still unexplored.

HISTORY: 1492–1865

In 1497, five years after Columbus landed in America, John Cabot sailed from England across the North Atlantic. Like Columbus, Cabot was looking for a route to Asia; like Columbus, he failed. Instead, he found the east coast of North America.

More than 100 years passed before the first English colony was established, at Jamestown, Virginia, in 1607. The colonists suffered hardships, but the growth of the tobacco trade made them prosperous. In 1619, Virginians began self-government in America with the first representative legislature.

The English Pilgrims fled from religious intolerance at home and landed at Plymouth, Massachusetts, in 1620. The Mayflower Compact of the Pilgrims also helped create a tradition of self-government. Other colonies—Maryland, Rhode Island, and Pennsylvania—were founded so that people of all faiths could practice their religion without persecution.

By 1754, 13 British colonies hugged the Atlantic coast. Then conflict between Great Britain (now the United Kingdom) and France in Europe spilled over into North America. Britain won the French and Indian War in 1763, and France lost Canada and all its North American land east of the Mississippi River except New Orleans.

The thirteen colonies stretched along the eastern coast of North America. French territory lay to the north and west of the colonies, and Spanish territory to the south.

The war for independence

Britain then tried to raise money by taxing the colonists and limiting their trade. However, the colonists argued that Britain had no right to tax them. The colonists expressed this belief in the slogan, "Taxation Without Representation Is Tyranny." Violent incidents like the Boston Massacre and acts of defiance like the Boston Tea Party followed.

Colonists and British soldiers finally clashed in 1775, the start of the American Revolution. In 1776, the colonists declared their independence from Britain and—with help from the French—defeated the British in 1781. In 1783, Britain recognized the United States of America.

The Constitutional Convention met to draw up the Constitution of the United States in 1787. George Washington (standing at the right) was president of the convention.

TIMELINE

1492	Columbus arrives in America.
1513	Ponce de León explores Florida.
1565	Spain founds St. Augustine, Florida.
1607	Jamestown colony founded.
1609	Henry Hudson reaches Hudson River.
1619	House of Burgesses, first representative legislature, established in Virginia.
1620	Pilgrims settle in New England.
1624	Dutch establish New Netherland colony.
1630	Boston founded by Puritans.
1636	Harvard College established.
1649	Maryland passes religious toleration act.
1664	British take over Dutch and Swedish colonies.
1681	Pennsylvania founded.
1718	France founds New Orleans.
1754–1763	French and Indian War.
1765	British pass Stamp Act.
1770	Boston Massacre.
1773	Boston Tea Party.
1775–1783	American Revolution.
1776	Declaration of Independence.
1781	British defeated at Yorktown.
1787	Constitution written.
1793	Eli Whitney invents cotton gin.
1800	Library of Congress established.
1803	Louisiana Purchase.
1804	Lewis and Clark explore the West.
1812–1814	War with British.
1819	Spain cedes Florida to United States.
1823	Monroe Doctrine set forth.
1825	Erie Canal opened.
1845	Annexation of Texas.
1846	British cede Oregon Country.
1846–1848	Mexican War; Southwest ceded.
1848	Gold discovered in California.
1853	Gadsden Purchase from Mexico.
1861–1865	American Civil War.
1863	Emancipation Proclamation frees slaves.
1865	President Lincoln assassinated.

George Washington (1732-1799)

Thomas Jefferson (1743-1826)

Benjamin Franklin (1706-1790)

A young and growing country

The founders of the United States—men like George Washington, Thomas Jefferson, and Benjamin Franklin—met to draw up a plan for a new government. They argued over how powerful this government should be. After one unsuccessful plan, the Constitution was written in 1787. The Constitution remains the law of the land to this day.

George Washington was elected the first president of the United States in 1789, and the country started to develop and prosper. Britain had given land east of the Mississippi to the United States. When Jefferson, as president, made the Louisiana Purchase from France in 1803, a vast area of land west of the Mississippi was added to the country. The annexation of Texas in 1845, war with Mexico in 1848, and agreements with Spain and Britain added even more land.

By the mid-1800's, the United States stretched from the Atlantic to the Pacific. During the early and mid-1800's, thousands of pioneers spread across the Western frontier. The settlement of the West led to many wars with Native Americans, who were driven from their homelands.

A country divided

Also by the mid-1800's, the dispute over slavery between the North and the South threatened the Union. In 1860 and 1861, 11 Southern states withdrew and formed the Confederacy. President Abraham Lincoln declared that the Union must be saved, and the American Civil War broke out. Lincoln freed the slaves in 1863. After a bitterly fought war, the North defeated the South in 1865, and the country remained united.

Southern troops attacked Fort Sumter, off Charleston, South Carolina, on April 12, 1861, starting the American Civil War.

HISTORY: 1866–PRESENT

The Civil War nearly ruined the South. During *Reconstruction,* a 12-year period following the war, plans were made to repair the region and return it to the Union.

Reconstruction laws gave rights to former slaves, and public schools were set up in the South. But many Southerners deeply resented Reconstruction. In time, they regained control of their state governments and made new laws that denied former slaves and their descendants their rights. For many years, the South remained a poor, agricultural region.

Before the war, the North had been industrially and financially stronger than the South. But the typical American business was still small. After the war, however, industry changed dramatically.

Machines replaced hand labor as the major means of production. A nationwide network of railroads provided easy transportation. New products were manufactured in large quantities. Big business grew, and the United States became an industrial giant.

The growing business activity attracted people from the country, who moved to the cities in record numbers. In addition, waves of immigrants came from other nations to work in U.S. factories and mines.

In the West, meanwhile, homesteaders settled and farmed the Plains, miners flocked to boom towns, and ranchers spread throughout the Southwest. Native Americans were gradually forced onto reservations.

War and hard times

Industrialization brought wealth to a few powerful businessmen, but most U.S. workers and farmers did not share in that wealth. Workers toiled long hours for little pay. Reformers tried to improve working conditions with new laws, and unions organized to fight for workers' rights.

The United States also began exercising its political and military power in Cuba and Central America. Nevertheless, when World War I (1914–1918) broke out in Europe, many Americans did not want to join the fight. By 1917, however, German attacks on U.S. ships led President Woodrow Wilson to declare war. By 1918, the United States had helped the Allies win World War I.

American astronaut Buzz Aldrin is photographed by fellow astronaut Neil Armstrong on the moon. Aldrin and Armstrong landed a spaceship on the moon on July 20, 1969. They planted an American flag on the lunar surface.

The economy seemed strong in the 1920's, but it was actually on shaky ground. The 1930's brought the country's worst economic slump ever—the Great Depression. Many people lost their jobs and their property. President Franklin Roosevelt's New Deal programs offered some relief.

World War II (1939–1945) was the most destructive conflict in human history. The United States entered the war against Nazi Germany and Japan in 1941, after Japan bombed the U.S. fleet at Pearl Harbor, Hawaii. The United States ended the war with Japan when it dropped two atomic bombs on Hiroshima and Nagasaki in 1945.

Postwar America

After the war, the United States was the leader of the non-Communist world. Over the next decades, it fought a *Cold War*—one without large-scale fighting—with the Soviet Union, the leader of the Communist world.

The 1960's were a time of change and crisis. The civil rights movement gained power for African Americans, but riots and crime erupted in poor city neighborhoods. Students and others protested the country's involvement in the Vietnam War (1957–1975).

In 1991, the United States led a United Nations coalition in air attacks against Iraq and its forces, which had been occupying neighboring Kuwait. After a short ground offensive, the coalition liberated Kuwait.

TIMELINE

1867	Alaska purchased from Russia.
1869	Transcontinental railway completed.
1871	Chicago Fire.
1876	Battle of Little Bighorn.
1876	Alexander Graham Bell invents telephone.
1879	Thomas Edison invents electric light.
1884	First skyscraper begun in Chicago.
1886	American Federation of Labor founded.
1898	Spanish-American War.
1898	Annexation of Hawaii.
1906	San Francisco earthquake.
1917	United States enters World War I.
1920	Women gain right to vote.
1920–1933	Prohibition (sale of alcohol banned).
1929	Stock market crash.
1933	Roosevelt's New Deal recovery plan.
1941	Japan attacks Pearl Harbor; United States enters World War II.
1944	Allied invasion of Europe launched under command of General Dwight D. Eisenhower.
1945	United States drops atomic bombs on Japan.
1945	United States joins the United Nations.
1950–1953	Korean War.
1961	Alan Shepard becomes first American in space.
1962	Cuban missile crisis.
1963	Nuclear test ban treaty signed.
1963	President Kennedy assassinated.
1964	Civil Rights Act enacted.
1965	United States enters the Vietnam War.
1968	Martin Luther King, Jr., assassinated.
1969	U.S. astronaut lands on moon.
1974	President Nixon resigns.
1980	Mount St. Helens erupts.
1981	American hostages in Iran freed.
1986	Space shuttle Challenger disaster.
1988	Nuclear arms treaty in effect.
1991	U.S.-led coalition liberates Kuwait from Iraqi forces.
1994	North American Free Trade Agreement takes effect.
1999	U.S. Senate acquits Bill Clinton of perjury and obstruction of justice.
2000	Ballot recounts and court challenges delay the outcome of presidential election.
2001	Terrorists hijack four U.S. airliners; two planes destroy World Trade Center, one slams into Pentagon, one crashes in Pennsylvania.
2003	Space shuttle Columbia disaster; United States launches military action against Iraq.
2009	Barack Obama becomes first African American president.
2011	War in Iraq ends.

Franklin D. Roosevelt (1882-1945)

Dwight D. Eisenhower (1890-1969)

Barack Obama (1961-)

Terrorist attacks and military action

On Sept. 11, 2001, about 3,000 people were killed in a violent terrorist attack. Terrorists hijacked commercial jetliners and deliberately crashed them into the World Trade Center in New York City and the Pentagon Building near Washington, D.C. Another hijacked jet crashed in Pennsylvania. Evidence pointed to al-Qa`ida, a terrorist organization that was sheltered in Afghanistan. The United States and its allies launched a military campaign against Afghanistan, and the campaign overthrew the Afghan government.

In 2003, the United States launched military action against Iraq. President George W. Bush said that Iraq's government posed a threat to the United States and other countries. The U.S. invasion led to the fall of the Iraqi regime. However, in the years that followed, establishing stability in Iraq and Afghanistan proved to be difficult. The war in Iraq officially ended in 2011.

In 2009, Barack Obama, a Democrat from Illinois, became the nation's first African American president.

The twin towers of the World Trade Center in New York City erupt in fire and smoke after terrorists crash hijacked airliners into them on Sept. 11, 2001. Approximately 3,000 people died in New York; in another hijacked plane attack at the Defense Department headquarters in Arlington, Virginia; and in the crash of a fourth hijacked jet in western Pennsylvania.

A MULTIETHNIC SOCIETY

The United States has long been called a "nation of immigrants." No country has received more immigrants; no country has ever attracted people from so many other nations. With a few exceptions—particularly American Indians, Alaskan Aleuts, and native Hawaiians—almost every United States resident has some immigrant ancestors. The nation's largest ethnic groups include people of English, German, Irish, French, Italian, Scottish, Polish, and Mexican descent, along with African Americans and American Indians.

Most white Americans trace their ancestry to Europe. Most black Americans are descendants of Africans who were forcibly brought to the United States as slaves. Most Hispanic Americans are immigrants or have immigrant ancestors who came from Latin America. A small percentage trace their ancestry directly back to Spain. Most Asian Americans trace their ancestry to China, the Philippines, India, Vietnam, Korea, or Japan.

Americans of many ethnic backgrounds gather on the National Mall in Washington., D.C., to celebrate Independence Day on the 4th of July. The Washington Monument rises in the background.

Waves of newcomers

The United States has had four great waves of immigrants. The first wave began in the 1600's, when immigrants from Europe settled on the east coast. Most were from England and other European countries.

The second wave of immigrants started in the 1820's and lasted for about fifty years. About 7-1/2 million newcomers arrived, nearly all from northern and western Europe. Many Irish people settled in the East, while Germans came to farm in the Midwest. Some Chinese came to find work on the railroads and in the mines when gold was discovered.

The third wave of immigrants was the largest. From the 1880's to 1920, about 23-1/2 million people poured into the United States. After 1890, the majority were from southern and eastern Europe.

Immigrants from Europe arrive at Ellis island in New York City. Beginning in 1892, millions of immigrants were processed through Ellis Island as they entered the United States.

The fourth wave came after 1965, when immigration from Asia and the West Indies rose dramatically. Many were refugees fleeing from war, persecution, or famine, like the immigrants from Southeast Asia in the 1970's. Others came in search of a higher standard of living. Today, most immigrants come from China, Colombia, Cuba, the Dominican Republic, El Salvador, India, Mexico, the Philippines, South Korea, and Vietnam. Far more immigrants come from Mexico than from any other country.

The American people

Some people call the United States a "melting pot." That term suggests a place where people from many lands have formed one unified culture. Americans have many things in common. Most speak English, wear basically the same kinds of clothing, and enjoy many of the same foods. Education, television, and radio have all helped shape an American identity.

However, many things now considered central to American life were actually introduced by immigrants, or created by them after they arrived. For example, jazz music developed as a mixture of elements from West African and European music, African American work songs, and spirituals. These last two forms of music

New York City's Chinatown is an example of the ethnic neighborhoods found in many large U.S. cities. Here, Chinese celebrate the start of their new year with a parade, fireworks, and long paper dragons. San Francisco, where large numbers of Chinese also settled, has its own Chinatown—a popular tourist attraction.

also developed into blues music and, later, rock music. Similarly, favorite foods like hamburgers and hot dogs are German in origin, while pizza is Italian.

In other ways, the United States is not a melting pot but rather a *culturally pluralistic* society—that is, a society where large numbers of people keep part of the culture and traditions of their ancestors. Many Americans take special pride in their origins. They consider themselves Irish Americans or Mexican Americans or Greek Americans, and they keep their own customs and traditions. Many cities have ethnic neighborhoods, where people of one particular national or ethnic background live. Ethnic restaurants, festivals, and parades reflect this cultural pluralism.

Some Americans have not accepted the idea of cultural pluralism, however. During the waves of immigration, some tried to stop the flow of newcomers. Discrimination and prejudice against certain groups have caused many problems. In an attempt to provide justice for all, Americans passed laws declaring that no one can be denied their rights on account of race, religion, color, or ethnic origin.

An Italian delicatessen brings a taste of southern Europe to the United States. People with Italian ancestry make up one of the largest ethnic groups in the country.

NATIVE AMERICANS

Native Americans lived in what is now America for thousands of years before Europeans came. They had no single name for themselves as a people, but almost every Native American group had its own name—one that represented the pride the group had in its way of life.

The Native American tribes of eastern North America lived in small villages. The Algonquin lived in domed wigwams. The Iroquois built long houses. These eastern groups hunted, fished, and grew such crops as corn, beans, and squash.

The Plains tribes lived in villages of tipis along rivers, where land was fertile. The women tended crops, while the men hunted deer, elk, and buffalo. After the 1500's, when Spanish explorers brought horses from Europe, the Plains tribes rode horses into battle and on buffalo hunts.

The Pueblo tribes of the Southwest farmed along rivers where they could irrigate their crops. They built large, many-storied homes of adobe and rocks. Other Southwest tribes moved about in small bands in search of food. These groups—the Apache and Navajo—often attacked the Pueblo people.

In the far Northwest, the Native Americans could catch plenty of fish, hunt game, and gather berries. They built plank houses with large wooden posts and beams, and carved totem poles depicting mythical beings.

California had a mild climate and an abundance of food. Modoc, Pomo, Maidu, and other California tribes lived in small villages and gathered wild plants, seeds, and nuts, especially acorns. They also hunted and fished.

The tribes that lived in the Great Basin east of California, however, had to adapt to a much drier climate. These people moved about in small bands searching for food.

Native Americans and Europeans

When Europeans first came to America's shores, most of the native peoples they met there were friendly. They taught the Europeans how to plant food and travel by canoe. The Europeans brought useful things for the Native Americans too, such as guns and horses. But the two groups had widely different ways of life. Soon, the cultures began to clash and small battles became common.

A Navajo shepherd guards her flock on the vast expanse of the Navajo reservation. The reservation lies in three states—Arizona, New Mexico, and Utah.

Native American leader Russell Jim (left) worked to block construction of a nuclear waste dump in the state of Washington. Many Native Americans have become active in social and political activities.

As the European settlers moved westward, they took land from the Native Americans. Some battles grew into wars. The settlers made numerous treaties with them, but quickly broke most of them. In 1830, the U.S. government began forcing all the eastern tribes to move west of the Mississippi.

The Indians of North America formed hundreds of tribes with many different ways of life. Scholars organize similar tribes into culture areas. Each culture area is shown in a different color.

A Mohawk construction worker treads a steel beam high above New York City. Since the late 1800's, many Mohawk have specialized in construction work on skyscrapers and bridges.

Navajo artists draw on the rich cultural heritage of the Southwest, keeping alive traditional skills in the crafting of locally mined silver and turquoise into beautiful jewelry.

The battles and bloodshed continued, though, once settlers themselves began to move west. Many members of the California tribes died from diseases brought by the Europeans or were murdered by gold miners. By killing off almost all the buffalo, the whites robbed the Plains tribes of their way of life. The U.S. Army fought the Sioux and Cheyenne frequently, often unjustly. Many tribes were moved onto reservations. The Native Americans in the Southwest and Northwest met the same fate.

Native Americans today

Today, about 3 million Native Americans, also known as American Indians, live in the United States. Native Americans, like all U.S. citizens, can live wherever they wish today, but many choose to make their homes on reservations. There, they can practice and preserve their traditional ways of life.

Native Americans were granted full citizenship and voting rights in 1924. The Bureau of Indian Affairs is supposed to promote the welfare of Native Americans, especially those living on reservations. In recent years, however, Native Americans have taken more control over their lands, schools, and other resources. Some Native Americans have sued the U.S. government over broken treaties and land that was illegally taken.

HISPANIC AMERICANS

In the years after Columbus landed in America, Spanish explorers, priests, and settlers created a region of Hispanic culture that spread from South America to what is now the Southern United States. Florida and the Southwest became part of the United States long ago, but much of the Hispanic influence remains today.

Today, about 50 million people of Hispanic descent live in the United States. They make up the nation's largest and fastest-growing minority group. Hispanic Americans have a high birth rate, and many continue to immigrate to the United States, especially from Mexico and the Caribbean.

Hispanic Americans have various national origins. The three largest Hispanic groups in the United States are Mexican Americans (more than 60 percent of the Hispanic population), Puerto Ricans (about 10 percent), and Cuban Americans and Dominican Americans (each about 3 percent). Others emigrated—or had ancestors who emigrated—from Central or South America or from Spain.

This colorful mural is part of a graffiti cover-up project. Murals that celebrate Hispanic American ethnic pride and cultural history are common in cities with large Hispanic populations.

Hispanic Americans also have various racial origins —a blending of white European, Native American, and black African. Most Mexican Americans are *mestizos*—people of mixed European and Native American ancestry. Many Puerto Ricans are of mixed Spanish and African descent, while others have some Native American ancestry as well.

Hispanic communities

About 90 percent of Hispanic Americans live in urban areas. Mexican Americans form the largest Hispanic group in most Southwestern cities, such as Los Angeles and San Antonio. Los Angeles also has small communities of Cubans, Guatemalans, and Puerto Ricans.

Many Hispanic leaders support the hiring of more Hispanic teachers for Spanish-speaking students. Bilingual education programs help students who do not speak English as their first language.

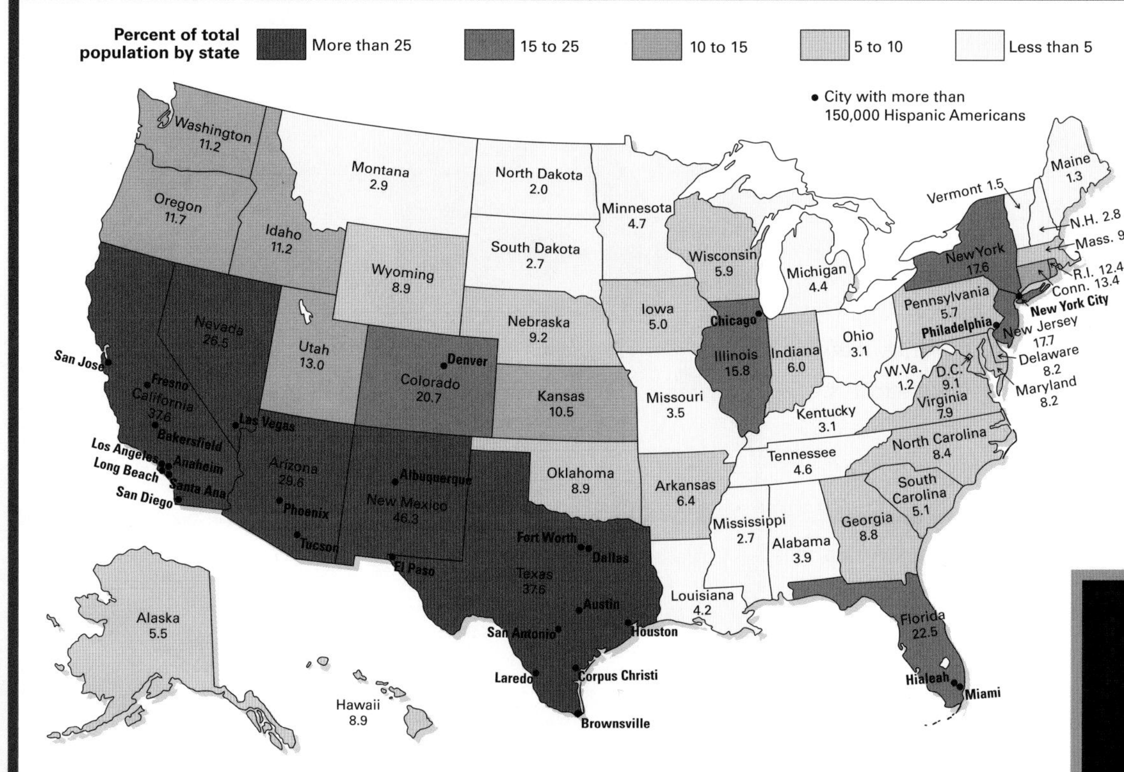

This map shows the state-by-state distribution of the U.S. Hispanic population, according to the 2010 census. Numbers on the map indicate the percentage of Hispanics in the total population of each state. The cities on the map have more than 150,000 Hispanic residents.

Miami, Florida, has the largest Cuban American population among U.S. cities. Beginning in the late 1950's, many Cubans came to the United States in order to escape the Communist government in Cuba.

A few Northern cities also have sizable Hispanic populations. About half of all Puerto Ricans in the mainland United States live in New York City. New York also has communities of people from Colombia, Cuba, the Dominican Republic, and Ecuador. Chicago has large groups of Mexican Americans, Puerto Ricans, and Cuban Americans. In recent years, Hispanic immigrants have begun to settle in smaller Northern cities too.

Hispanic culture

Most Hispanic Americans speak English, but they use Spanish as well, especially at home. Many are proud of their heritage and feel they should not lose touch with it. They want to be bicultural as well as bilingual.

Hispanic Americans place a high value on family life. Religion is also a common cultural heritage among Hispanics. A large majority practice the Roman Catholic religion brought by the Spanish missionaries.

The Hispanic influence on American culture can be seen in many areas, including food, music, and art. For example, tacos and enchiladas are popular Mexican foods. Latin and Cuban styles of music have "crossed over" into popular American culture.

In 2009, Sonia Sotomayor became the first Hispanic American to be appointed a justice of the Supreme Court of the United States.

Like other minorities, Hispanic Americans have suffered from discrimination in jobs, housing, and education. Some Hispanics are also hampered by not having skills that are important for competing in U.S. society. For example, many new Hispanic immigrants cannot speak or understand English. Discrimination and the lack of such skills have contributed to a high rate of unemployment—and, consequently, a high rate of poverty—among Hispanics. Although millions of Hispanic Americans have overcome these obstacles, many others remain in poverty.

AFRICAN AMERICANS

The 40 million African Americans living in the United States today make up the country's second largest minority group, after Hispanic Americans. Most African Americans trace their origins to an area in western Africa that was controlled by three great and wealthy empires—the Ghana, Mali, and Songhai.

In the early 1500's, Europeans joined in the slave trade that had been carried on in Africa since ancient times. Over the next 300 years, millions of black Africans were enslaved and shipped to colonies in America. Until 1863, most of these Africans worked as slaves throughout the South.

Slavery was officially abolished in 1865, and the slaves were freed. In 1868, the 14th Amendment gave African Americans equal rights. In 1870, the 15th Amendment stated that African American men could not be denied the right to vote because of their color. Reconstruction in the South also tried to improve the political power of the freed slaves.

The progress African Americans made during these years was not allowed to last, however. Southern whites began to pass laws that discriminated against them. African Americans living in the Southern States increasingly lost their voting rights and were segregated (kept apart from whites) in such public facilities as schools, buses, and restaurants. In 1896, the Supreme Court of the United States ruled that such segregation was legal as long as the public facilities were "equal."

Secretary of State Colin Powell and National Security Adviser Condoleezza Rice became the first African Americans to hold these posts. Both were appointed by President George W. Bush in 2001. Rice became secretary of state in 2005.

Voter registration has been a chief goal among many African American leaders. These leaders stress the importance of political participation in helping solve the problems that face African Americans today.

In the 1900's, African Americans continued to suffer from discrimination, and they had little opportunity to better their lives. Even those who moved to the North did not find answers to their problems. There, too, they were segregated, not by law but in practice. Because they were poor and poorly educated, African Americans had to live in run-down areas of the large cities. These poor communities developed into slum ghettos.

The civil rights movement

In 1954, a historic ruling was handed down by the *Supreme Court*. In *Brown v. Board of Education of Topeka*, the high court reversed the 1896 decision that "separate but equal" public schools were legal. The Supreme Court justices ruled that an African American student, Linda Brown, should be allowed to attend an all-white school near her home.

African Americans began to strike out against discrimination. In Montgomery, Alabama, in 1955, an African American woman named Rosa Parks was arrested for refusing to give up her seat on a bus to a white person. A city law required African Americans to do so. African Americans boycotted the buses by refusing to ride in them until the city abolished the law. The boycott cast the nation's attention on its leader, Martin Luther King, Jr., a Baptist minister.

With other civil rights leaders, King went on to lead boycotts and protest marches in the South and then the North. Riots in ghettos broke out during the 1960's, but King urged African Americans to use only peaceful means to reach their goals.

King was assassinated in 1968. Still, civil rights laws passed during the 1960's banned discrimination in voting, jobs, housing, and public places like hotels and restaurants.

A black female architect discusses the design of a project with three colleagues. African Americans—women and men—have made progress since the 1960's in professions previously closed to them.

African Americans today

African Americans have made great progress since the 1960's. Gains in education have been important. Although the dropout rate remains high, African American enrollment in high school and college has increased.

The number of African American-owned businesses has increased also, and the African American middle class is growing. Nevertheless, a large number of African American families still live in poverty, and the rate of unemployment is much higher among African Americans than among whites.

Sports, entertainment, and the arts have been open to African Americans longer than other fields, and many African Americans have made great achievements in these areas. Baseball player Hank Aaron, football player Walter Payton, basketball player Michael Jordan, track and field star Jackie Joyner-Kersee, and golfer Tiger Woods all set records in their sports. Entertainers like Oprah Winfrey and Queen Latifah became very popular. Novelists Alice Walker and Toni Morrison wrote award-winning books.

African Americans also have played important roles in government, politics, and the military. African Americans who have served as mayors of major U.S. cities include Thomas Bradley of Los Angeles, David Dinkins of New York City, and Harold Washington of Chicago. Andrew Young served as U.S. ambassador to the United Nations as well as mayor of Atlanta. In 1990, L. Douglas Wilder took office as governor of Virginia—the country's first elected African American governor. In 1984 and 1988, Jesse Jackson waged strong campaigns to become a candidate for president; in the 1990's, Jackson advised President Bill Clinton on human and civil rights issues. In 1983, Guion Stewart Bluford, Jr., became the first African American to travel in space. Colin L. Powell became the first African American chairman of the Joint Chiefs of Staff in 1989 and secretary of state in 2001. Also in 2001, Condoleezza Rice became the first African American National Security Adviser.

In 2009, Barack Obama became the first African American president. Obama, a Democrat, had served as a senator from Illinois before being elected to the presidency. Also that year, Eric Holder became the first African American attorney general.

URBAN LIFE

In 1790, when the first official U.S. census was taken, about 95 percent of the nation's people lived in rural areas. Only 5 percent lived in cities like Philadelphia, New York, and Boston.

Through the years, however, a dramatic shift occurred. When agricultural methods and equipment improved, farming became more efficient. From the 1800's on, far fewer workers were needed on farms. About this time, the industrial boom created numerous factory jobs in urban areas. As a result, a steady flow of people moved from rural areas to cities to find work.

The fast-paced life of cities appealed to many rural people, especially younger ones. Large numbers of them left rural America for the excitement of the "big city."

The wave of immigrants to the United States from the 1890's to 1920 also swelled city populations. Many of these immigrants settled in tenements (crowded, decaying apartment buildings) in New York City and other urban areas.

Urban areas—ranging in size from small towns to huge cities—dot the U.S. landscape. Today, they take up only about 3 percent of the land, but they house about four-fifths of the nation's people.

People in New York City, and a few other American cities, live in close quarters in apartment buildings. Even people with middle class incomes live in small spaces that command high rents.

A jogger runs along the East River on the Upper East Side of New York City. Open spaces and oasis of green provide a respite from the congestion of urban life.

Most cities are not isolated communities but instead are surrounded by suburbs. Suburbs grew rapidly during the mid-1900's, as many people moved to these outlying areas and commuted to work in the cities. Automobiles contributed greatly to the growth of suburbs, because they provided more convenient transportation for people between the cities and neighboring areas.

The move to the suburbs was so massive that in 1980, for the first time, more Americans lived in suburbs than in cities. Now, many suburban residents work and seek entertainment in the suburbs instead of traveling to the city.

A *metropolitan* area consists of a city and the surrounding developed area. There are hundreds of metropolitan areas in the United States. The three largest are, in order of size, the New York-Northern New Jersey-Long Island, Los Angeles-Long Beach-Santa Ana, and Chicago-Naperville-Joliet areas.

Urban areas offer a wide variety of jobs, ranging from medical personnel and office workers to trash collectors and transportation workers. Cities also offer special services, shops, night life, and cultural events that rural areas often lack. Concerts, art galleries, theaters, sporting events, and museums help make cities exciting. The variety of cultural backgrounds also makes urban areas interesting. Large cities often have a number of ethnic neighborhoods.

Large cities are often divided economically. Urban society includes extremely wealthy and extremely poor people, as well as a huge middle class. Wealthy people live in large, luxurious apartments, condominiums, or single-family homes. The large urban middle class lives in similar but more modest housing. Most poor people live in small, crowded apartments or run-down single-family homes. Sometimes public housing is available, paid for in part by the government.

Poverty and substandard housing are just two of the many problems facing U.S. cities. Crime, noise, air pollution, and traffic congestion all come with the advantages urban areas in the United States offer.

In this leafy suburb of Washington, D.C., large single-family homes and well-groomed yards are common.

Suburban housing subdivisions lie just outside big cities in the United States, like this community near Houston. Residents may work in the city but they live in the suburbs because there is more space and the school system may be better than the city system.

THE MISSISSIPPI RIVER

The Mississippi is the second longest river in the United States. Only the Missouri is longer. The Mississippi begins as a small, clear stream emerging from Lake Itasca in northwest Minnesota, and it then flows 2,340 miles (3,766 kilometers) to the Gulf of Mexico. As it flows, the Mississippi and its tributaries drain almost all the plains that lie between the Appalachian Mountains and the Rocky Mountains. This river basin includes the nation's most productive agricultural and industrial regions—more than 1 million square miles (3 million square kilometers) of land.

The Native Americans who lived in the upper Mississippi Valley chose the name for the river. They called it Mississippi, meaning big river. Today, "Old Man River"—as the mighty waterway is sometimes called—forms part of the boundary of 10 states.

The Illinois and Missouri rivers are two of the major northern tributaries of the Mississippi. The Missouri River is muddy, and where it meets the Mississippi River, near St. Louis, its waters begin to turn the Mississippi muddy as well. In the South, the Mississippi is known for this muddy color.

The Ohio River, flowing into the Mississippi River at Cairo, Illinois, doubles the amount of water in the river. This spot divides the upper Mississippi from the lower Mississippi. The flood plain of the lower Mississippi forms a broad, fertile valley where the river twists and turns in wide loops.

The Mississippi River is the second longest in the United States. Only the Missouri River is longer. The Mississippi flows 2,340 miles (3,766 kilometers) from its source in northwestern Minnesota to its mouth in the Gulf of Mexico.

The Mississippi River drains the heart of the North American continent, carrying vast quanities of sediment, which is deposited along the shore of Louisiana and into the Gulf of Mexico.

The Mississippi River supports a rich variety of wildlife—animals, birds, and insects—as well as plant life.

By the time the Mississippi nears its mouth in the Gulf of Mexico, it is carrying large amounts of silt (soil particles). The river deposits this fertile soil to form a huge delta. The Mississippi Delta is about 13,000 square miles (33,700 square kilometers).

The Mississippi has played a key role in U.S. history. In 1682, the French explorer Sieur de La Salle traveled down the river to its mouth. He claimed the river and all the land it drained for the king of France. Later, France lost the land east of the Mississippi to Great Britain. In 1803, the French sold land west of the river to the United States in the Louisiana Purchase. After the entire Mississippi Valley became part of the United States, settlers and traders set out on the river in flatboats, keelboats, and rafts. In the 1880's, steamboats turned the river into the great transportation and trade route it is today.

The river today carries about 500 million short tons (450 million metric tons) of freight every year—about half of all the freight carried on the nation's inland waterways. River barges pushed by tugboats carry most of this freight.

Between Minneapolis and Cairo, the southbound freight consists mainly of agricultural products, such as corn, soybeans, and wheat. Coal and steel products from the Ohio River system are transported north. South of Cairo, goods from the Ohio double the Mississippi's traffic. Most of the cargo consists of southbound agricultural goods, coal, and steel products.

At Baton Rouge, Louisiana, petrochemical products, aluminum, and petroleum are added to the barge traffic. Beginning at Baton Rouge, the Mississippi deepens and allows passage of ocean-going vessels. The greatest volume of traffic on the Mississippi moves between New Orleans and the gulf.

The Gateway Arch in St. Louis, Missouri, represents the city's historic location as the gateway to the West. St. Louis is the busiest inland port on the Mississippi River.

ENERGY AND RESOURCES

In their homes and offices, cars and trucks, and farms and factories, the people of the United States use a vast amount of energy. Today, the United States has only about 5 percent of the world's population, but the nation consumes about 20 percent of the world's energy.

Energy comes from various sources. Petroleum, or oil, provides about 35 percent of the energy used in the United States. It is used to power cars, ships, and airplanes, and it heats millions of homes, offices, and factories.

Natural gas produces about 25 percent of the energy used nationwide. Many industries burn natural gas for heat and power, and millions of people use it to heat their homes, cook their food, and dry their laundry.

Coal, which supplies about 20 percent of U.S. energy, is used mainly in the production of electricity and steel. The electricity in turn lights buildings and powers machinery for offices, factories, and farms.

Glen Canyon Dam in Arizona harnesses the Colorado River to produce electricity for businesses and homes. Water power helps meet the country's high level of energy consumption.

Water produces less than 5 percent of America's energy, mainly in hydroelectric plants where electricity is produced for industries and homes. Nuclear power produces about 10 percent of the nation's energy. Like water, nuclear power generates electricity for industries and homes.

The United States is a major producer of energy resources. It is third in the world in the production of petroleum and natural gas (after Russia and Saudi Arabia), and second in the production of coal (after China). The United States has many other natural resources as well.

Potash a salt mainly used to make fertilizer, can be extracted from evaporation ponds. Most of the potash produced in the United States comes from New Mexico.

A train carries coils of steel from a mill in East Chicago, Indiana. Steel production requires two resources that are plentiful in the United States—coal and iron ore.

Besides the minerals used to produce energy, it has valuable deposits of copper, gold, iron ore, lead, phosphates, potash, silver, sulfur, and zinc.

The rich soils of the land are another major resource. The most fertile lands include the dark soils of the Midwest and the alluvial (water-deposited) soils along the lower Mississippi River and in other river valleys.

Fresh water—in the country's lakes and rivers and under the ground—is also a precious natural resource. About 400 billion gallons (1,500 billion liters) of water is used every day in the United States, mainly to operate manufacturing and power plants, and to irrigate farmland. Homes use about 10 percent of the total.

Fish in the coastal waters of the country are another valuable resource. The greatest quantities are taken from the Pacific Ocean, which supplies cod, crabs, halibut, pollock, salmon, tuna, and other fish. Leading catches from the Gulf of Mexico include crabs, menhaden, oysters, and shrimp. The Atlantic yields flounder, herring, menhaden, and other fish, and such shellfish as clams, crabs, lobsters, oysters, and scallops.

Finally, the forests that cover almost one-third of the United States yield many valuable products. The forests of the Pacific Northwest supply about a fourth of all the nation's lumber. Southern forests also provide lumber, as well as wood pulp for making paper, turpentine, pitch, rosin, and wood tar. The Appalachian forests and areas around the Great Lakes produce fine hardwoods, such as hickory, maple, and oak, for making furniture.

The moderate climate enjoyed by much of the United States can also be considered a natural resource. It has allowed people to settle in most parts of the country and enabled farmers to grow a wide variety of crops.

Lumber is an important resource of the United States. To help conserve the country's forests, foresters replant areas where they have cut timber. They may also raise seedlings in a nursery, and transplant young trees to the forest.

INDUSTRY

Modern telephone and Internet inventions have revolutionized the communications industry. The most popular include the "smartphone" and the laptop computer.

The United States ranks first in the world in the total value of its economic production. The nation's *gross domestic product* (GDP)—the value of all the goods and services produced within a country in a year—is about three times as large as the gross domestic product of China, which ranks second.

The growing service economy

Service industries—a general name for a large, varied category of economic activities that produce services rather than goods—account for the largest portion of the U.S. GDP. Service industries include such areas as community, social, and personal services; finance; trade; government; and communications. A majority of the country's workers are employed in service industries.

Community, social, and personal services form the most important service industry in terms of percentage of the U.S. GDP. This category includes health care, legal services, nursery schools, and many other types of businesses.

The finance industry supports economic activity by supplying money in the form of loans from banks. Two other kinds of financial institutions are securities exchanges, where stocks and bonds are bought and sold to raise money for businesses, and commodities exchanges, where such goods as grains and precious metals are bought and sold.

Wholesale and retail trade, together with restaurants and hotels, play major roles in the American economy. In *wholesale trade*, a buyer purchases products directly from a producer and then sells the products to retailers. For example, a wholesaler for vegetables buys large amounts of vegetables from the growers and then sells them to grocers. In *retail trade*, goods are sold directly to the consumer. The grocers, who sell the vegetables, are in retail trade. Restaurants and hotels greatly benefit from the tens of millions of foreign tourists who visit the United States annually.

International trade provides markets for agricultural and manufactured goods produced in the United States. The nation imports goods that it lacks entirely or that producers do not supply in sufficient quantities. It also imports goods produced by foreign companies that compete with U.S. firms. Canada, China, Germany, Japan, and Mexico are some of the country's chief trading partners.

The government provides many services to the people of the country. Local and state governments and the federal government employ many American workers. They provide such services as police protection, education, and trash collection.

The communications industry includes publishing and broadcasting companies that provide news and entertainment. Telephone and Internet services and mail delivery are also part of the communications industry.

Manufacturing and construction

Manufacturing is also an important economic activity, both in terms of employment and the gross domestic product. The value of American manufactured goods is greater than that of any other country.

Factories in the United States turn out a wide variety of *producer goods* (articles used to make other products), such as sheet metal and printing presses. They also manufacture *consumer goods* (products that people use), such as cars, clothes, and television sets. The leading kinds of manufactured goods are chemicals, transportation equipment, food products, computer and electronic products, fabricated metal products, machinery, petroleum and coal products, and plastics and rubber products.

The Midwest and Northeast have long been centers of manufacturing. Midwestern factories produce much of the nation's iron and steel, as well as cars. The Northeast has many food-processing plants, printing plants, and manufacturers of electronic equipment.

Since the mid-1900's, the fastest-growing manufacturing areas have been on the West Coast, in the Southwest, and in the South. California produces aircraft, aerospace equipment, computers and electronic components, and food products. In Texas and other Gulf States, petroleum refineries and petrochemical industries are major manufacturers.

The construction industry also provides jobs for many U.S. workers. Architects, engineers, contractors, bricklayers, carpenters, and electricians build homes, offices, and factories across the country.

An operator works a computer-assisted crane to hoist the steel beams of a skyscraper under construction.

Workers assemble an automobile at a plant in Chicago. Automobile manufacturing has been a backbone of American industry for generations.

Rows of farm tractors await shipment on the docks of the Port of Baltimore. Modern farm equipment, a major U.S. export, has made agriculture more efficient and productive worldwide.

AGRICULTURE

The United States not only is a world leader in agricultural production, but it also helps feed the world. Every year, U.S. farmers produce more than enough food to feed all of the people in the United States, so the country also exports large amounts of food. Food exports account for about one-third of U.S. farm income.

The most valuable farm product is beef cattle, followed by corn, soybeans, dairy products, broilers (young, tender chickens), greenhouse and nursery products, hogs, wheat, and chicken eggs. Farms in the United States also produce large amounts of almonds, apples, cotton, grapes, hay, lettuce, potatoes, rice, tomatoes, and turkeys.

Farmers throughout the country raise dairy cattle for milk, cheese, and other products, but dairy production is especially concentrated in a northern zone running from Minnesota to New York. California is also a major dairying state. Millions of beef cattle are raised on huge ranches in the West, and large numbers are also found in the Midwest and South. The Midwest is noted for its hogs, and the South for its chickens.

The Midwest, the country's Corn Belt, also accounts for much of America's soybean production. The Wheat Belt stretches across the Great Plains. Almost all the country's cotton is grown in California, the Southwest, and the South. Farmers throughout the nation produce poultry, eggs, fruits, vegetables, nuts, and many other crops.

During the 1900's, the number of farms in the United States decreased, but the size of the average individual farm increased. The United States today has about 2 million farms, compared with about 6.5 million in the 1920's. The average size of a farm today is 420 acres (170 hectares), compared with about 143 acres (58 hectares) in the 1920's.

Farms in California produce hundreds of commodities, including chili peppers. Agriculture in California generates more income than agriculture in any other state.

This map shows the country's major agricultural areas. Cropland is concentrated in the middle of the United States. Most of the grazing land is in the West.

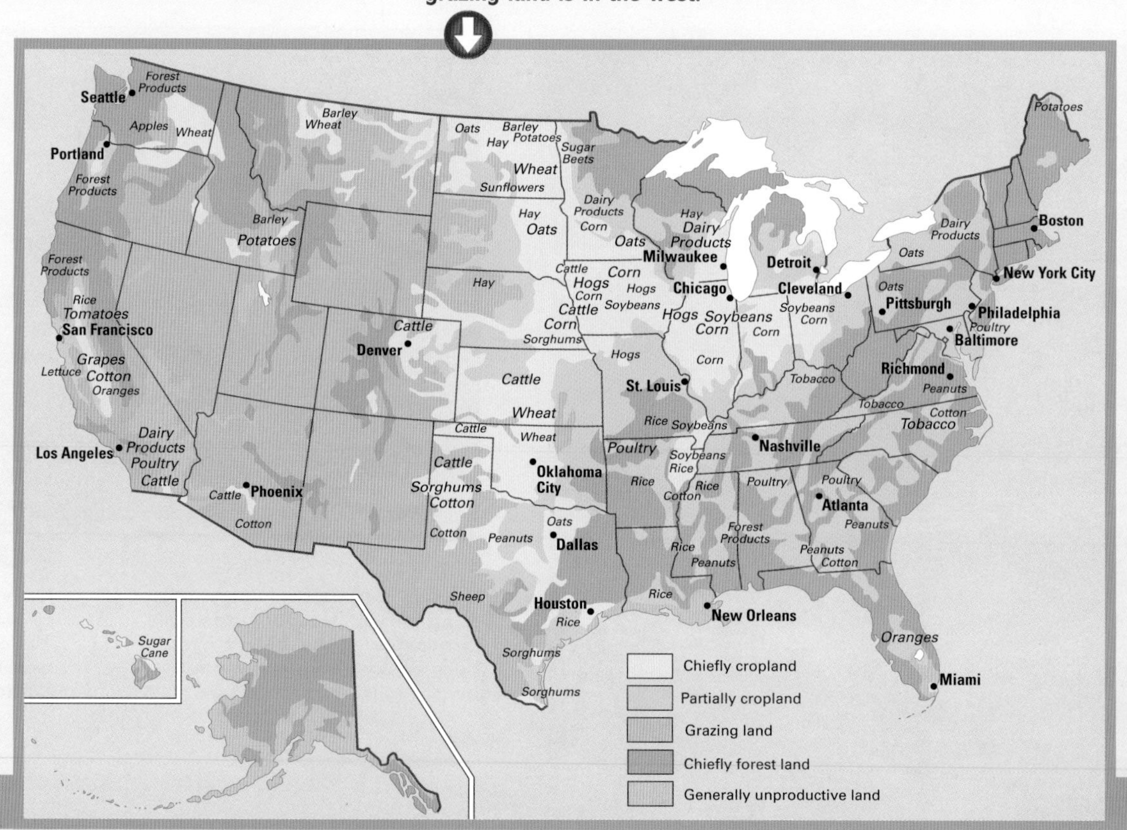

Kansas grain is loaded onto barges on the Missouri River. Kansas is a national leader in wheat production. Wheat is ground into flour to make such foods as bread and pasta.

Rows of corn cover much of the farmland in a region of the midwestern United States called the Corn Belt. Corn is one of the world's most important crops. It is a leading source of energy in the human diet, serves as a major livestock feed, and has many industrial applications.

America's farms produce more now than ever before, largely for two reasons: (1) the use of modern farm machines and agricultural methods and (2) efficient management. American inventors devised the harvesting machine and steel plow. American scientists have helped develop improved livestock breeds and plant varieties, as well as chemicals for fertilizing soil and controlling weeds and insects.

In addition, farms are now managed much like other businesses. Good business management is essential because the costs of farming have increased tremendously. Today, the typical farm requires a start-up investment of hundreds of thousands of dollars. Most of the money goes for land. The rest goes mainly for supplies and equipment. Many farmers, unable to meet the rising costs of farming, have been forced to quit and sell their land.

The great majority of the nation's farms—more than 90 percent—are owned by individuals or by partnerships or corporations made up of family members. In many cases, the owners operate the farms, but some owners rent land to other farmers. More than half of all farmland in the United States is rented. Some of the largest farms in the country are owned by such corporations as food-processing companies or feed manufacturers, who hire managers to run them.

American farmers today lead very different lives from those of their grandparents. Machines, such as tractors, harvesters, and conveyor systems, have eliminated much back-breaking work. Today, the homes of farm families have most of the same comforts and conveniences as those of urban people.

A cattle rancher feeds hay to his stock in winter, when natural forage is scarce. The United States produces more beef than any other country. The U.S. cattle industry is centered in the Western states, but cattle are raised in many other states including Florida.

AMERICAN WAY OF LIFE

The Stars and Stripes—the U.S. flag—flies outside a house on Long Island, in New York.

The United States was built by freedom-loving people who came from different parts of the world. Through the years, they and their descendants learned to live and work together, and they became proud of being American. This cooperation and shared pride made the United States the powerful, diverse, and wealthy nation it is today.

The United States is in many ways a nation of cultural pluralism, where many people keep alive the traditions of their ancestors. But there is also an American culture—a style and a special way of looking at life that Americans share.

American values

Most immigrants came to the United States seeking political or religious freedom or economic opportunity. Today, Americans still value these ideals highly. Freedom of speech, freedom of the press, and freedom of religion are enjoyed by the people and guaranteed by the Constitution. *Civil rights*—equal treatment for all people, including equal opportunity in jobs—are guaranteed by laws.

Economic opportunity enables people in the United States to move from one social class to another. As in many other societies, the values of the middle class are the nation's most widely held values. Middle-class Americans stress self-improvement and economic success. They believe it is important to work hard, make a good living, and follow the community's standards of morality. They also value education, and many send their children to college.

About 80 percent of Americans are members of a religious group. Although every religious group is free to worship as it chooses, the Christian religion has a stronger influence on American life than any other faith. For example, most people do not work on Sunday, the Christian day of worship. Protestants make up the largest Christian group in the United States, followed by Roman Catholics. Buddhists, Hindus, Jews, Mormons, Muslims, and members of Eastern Orthodox Churches each make up a small percentage of the population.

Shopping malls are popular gathering places for Americans. Malls offer stores, restaurants, and entertainment outlets in one convenient location. The Mall of America in Bloomington, Minnesota, outside Minneapolis, is the largest combined retail and entertainment center in the United States.

Americans enjoy sports, both as spectators and as players. Football draws millions of people to stadiums each year to watch their favorite professional, college, and high school teams. Millions more watch football games on television.

The American family

Traditionally, the American family has consisted of a mother, a father, and several children. This kind of family structure is called a *nuclear family.*

Today, many people have turned away from this traditional family pattern. Now, on the average, married couples have only about two children. Some couples decide to have none. About half of the marriages that take place in the United States end in divorce. Divorced or widowed parents sometimes choose not to remarry. They and their children live together in single-parent families. The growing number of working wives and mothers has also changed the pattern of family life in many ways.

Although American family life changed greatly in the second half of the 20th century—as people adopted new roles and developed new kinds of family structures—Thanksgiving dinner remains a vibrant annual tradition in the United States.

Recreation

Many Americans have a great deal of leisure time, and they spend it in a variety of ways. They pursue hobbies; take part in sports activities; attend sporting and cultural events; watch movies and television; listen to music; read books and magazines; and communicate or find news and entertainment through the Internet. They also enjoy trips to museums, beaches, parks, playgrounds, and zoos. They take weekend and vacation trips, eat at restaurants, and entertain friends at home. These and other activities contribute to the richness and diversity of American life.

Hobbies occupy much of the leisure time of many Americans. Many people enjoy raising flower or vegetable gardens or indoor plants. Other popular hobbies include stamp collecting, coin collecting, photography, and writing blogs (personal journals on the Internet). Interest in such crafts as pottery making, quilting, weaving, and woodworking is also common.

ARTS AND ENTERTAINMENT

The first American artists were Native Americans. They used their skills in pottery, weaving, and carving to make everyday objects beautiful as well as useful. After the arrival of Europeans, some of the earliest major American works of art were the houses built by the colonists. *Colonial style* houses date back to the 1600's.

During the 1700's, American craftworkers began producing outstanding examples of furniture and silver work. John Singleton Copley and other American painters of the period created excellent portraits.

American literature first gained recognition in Europe in the early 1800's. Washington Irving combined the styles of the essay and the sketch to create a new literary form, the short story. Irving's "Rip Van Winkle" and "The Legend of Sleepy Hollow" are probably his best-loved works. James Fenimore Cooper's series of five novels, *The Leather-Stocking Tales,* contained the first serious portrayal of American frontier characters and scenes.

During the late 1800's, American architects began designing skyscrapers that revolutionized urban architecture throughout the world. William Le Baron Jenney designed the first metal-frame skyscraper, completed in Chicago in 1885.

The late 1800's and early 1900's saw the birth of two uniquely American art forms—jazz and musical comedy. Motion pictures and modern dance soon followed.

The Guggenheim Museum in New York City houses a major international art collection. Many consider the building itself, designed by Frank Lloyd Wright, the collection's most important work of art.

Painting

By the mid-1800's, many U.S. artists had come to feel that the country's landscape was the perfect subject for a truly American style of painting. New York's Hudson River Valley and the West inspired many artists.

The American public got its first look at modern art in 1913 at a famous exhibit called the Armory Show. Many American artists then adopted the European modern style. In the 1930's, some returned to American themes. Grant Wood painted Midwest scenes, and Edward Hopper depicted urban life.

Since World War II (1939–1945), the United States has largely replaced Europe as the center of Western painting. Modern painting has produced abstract art, like that of Jackson Pollock; realistic art, like that of Andrew Wyeth; and pop art, like that of Andy Warhol.

Dance and theater

Modern dance, which developed in the early 1900's, has centered on U.S. dancers and dance companies since the 1940's. America's famous modern choreographers include Twyla Tharp, Alvin Ailey, and Martha Graham, who formed her own dance company and created dances for ballets such as *Appalachian Spring.*

Today, theater groups all across the United States produce plays. Broadway in New York City is the best-known center of theater in the country, but cities around the nation have their own acting companies. Many towns have community theaters.

An outdoor concert at the Jay Pritzker Pavilion in Chicago attracts music lovers through the summer. The extravagant band shell, designed by American architect Frank Gehry, has made the pavilion a major tourist attraction.

George Gershwin combined elements of jazz with serious and popular music to produce musical comedies, popular songs, symphonic works, and opera. His *Rhapsody in Blue* is probably the best-known orchestral piece written by an American.

Country music, once the folk music of Southern whites, is now popular nationwide. Rock music is a mixture of blues, jazz, and American country music. Singers Elvis Presley and Chuck Berry helped make rock music the leading type of popular music since the 1950's. Since the late 1900's, rap and hip-hop music, which were pioneered by urban African Americans, have become among the most popular and influential forms of music in the United States.

Walt Disney World near Orlando, Florida, uses themes and characters from Disney's animated films to entertain millions each year.

American playwrights have produced masterpieces of serious drama as well as musical comedy. Arthur Miller's *Death of a Salesman* deals with a salesman who fails to realize the American dream of success; the play has been performed around the world. *A Chorus Line,* a musical about dancers, was one of the longest-running shows in the history of Broadway theater.

Movies and music

Motion pictures are one of the most popular and influential art forms in the United States, ranging from early silent films like *The Birth of a Nation* to later "blockbuster" movies such as *Gone with the Wind, Star Wars,* and *Avatar.* Walt Disney's *animated* (cartoon) films have been celebrated for their artistic merit as well as their entertainment value.

Popular music has taken many forms in the United States. Blues sprang from the songs of slaves. Jazz first became popular about 1900. Louis Armstrong, a trumpeter, was the first great jazz soloist.

Motion pictures are one of the most popular and influential art and entertainment forms in the United States. Warner Brothers in Burbank, California, is one of the world's leading motion picture studios.

URUGUAY

On the southeastern coast of South America lies the continent's second smallest independent nation—Uruguay. A land of gently rolling plains and beautiful sandy beaches, Uruguay is bordered by Argentina on the west, Brazil on the north and east, and the Río de la Plata and Atlantic Ocean on the south. Uruguay's developing economy relies heavily on agriculture, with farm products such as beef, milk, and wool providing most of the country's export income.

A cheerful waiter serves his customers in a crowded city cafe. Cafes provide lively social centers for middle-class city dwellers. Most of the people who live in Montevideo and other Uruguayan cities belong to the middle class.

Most Uruguayans are descended from Spanish settlers who came to the country in the 1600's and 1700's and Italian immigrants who arrived during the 1800's and early 1900's. Nearly all Uruguayans speak Spanish, the country's official language. It is generally spoken with an Italian accent. Many Uruguayans also speak a second language, usually English, French, or Italian.

A Uruguayan enjoys his *yerba maté* (tea)—the country's national beverage—in the traditional way, sipped through a silver straw from a gourd. This refreshing drink is made by pouring hot water onto the dried leaves of a holly tree.

Uruguay's people reflect a wide range of personalities and occupations. Free-spirited gauchos herd cattle on huge ranches in the interior of the country, while hard-working farmers grow barley, corn, potatoes, and other crops in the fertile valleys of the Uruguay River and the Río de la Plata. Urban dwellers in Montevideo, the nation's capital, hold government or professional jobs or work in business and industry.

In their leisure time, people crowd the shops and cafes along the city's treelined boulevards. In cities and towns throughout the country, young soccer fans play their favorite sport with boundless energy.

During the early 1900's, Uruguay developed into one of the most wealthy and democratic nations in South America. However, a decline in the economy in the 1950's and 1960's led to widespread political unrest, terrorism, and military rule. Today, Uruguay is once again ruled by an elected civilian government. However, many of the country's economic problems remain unsolved.

Early history

Long before the first Europeans arrived in what is now Uruguay, the region was inhabited by Indian tribes. The largest of these tribes were the Charrúas. In 1516, when the Spanish explorer Juan Díaz de Solís sailed into the Río de la Plata, he and his expedition were killed by the Charrúas.

When later Spanish explorers found that Uruguay lacked deposits of precious metals, the region was all but forgotten for more than a century. But in 1680, Portuguese soldiers from Brazil established the town of Nova Colonia do Sacramento (now Colonia) on the Río de la Plata, across from the Spanish settlement of Buenos Aires.

To prevent the Portuguese from expanding any farther into Spanish territory, the Spaniards founded the town of Montevideo in 1726. By the 1770's, the Spanish had settled most of Uruguay. In 1776, the region became part of the Spanish colony called the Viceroyalty of the Río de La Plata, also called the Viceroyalty of La Plata. The Portuguese formally gave up what is now Colonia in 1777.

A group of Uruguayan students enjoy a break during afternoon recess. The government provides free public schooling through the university level, and the law requires children from ages 6 through 15 to attend school.

Artigas and the Tupamaros

During the early 1800's, a Uruguayan soldier named José Gervasio Artigas organized an army to fight for independence from Spain. Just as Artigas's army was about to defeat the Spaniards at Montevideo, the Portuguese attacked both the Uruguayan and Spanish troops.

Rather than submit to either Spanish or Portuguese rule, Artigas led his army and thousands of Uruguayans across the border to Paraguay and Argentina. Spanish control over Uruguay ended in 1814, and in 1815 Artigas and his troops took control of Montevideo.

In 1816, Portuguese troops again attacked Uruguay. After four years of bitter fighting, the Portuguese annexed Uruguay to Brazil and forced Artigas into exile.

In 1825, the battle for Uruguayan independence was renewed under the leadership of a group of patriots called "The Immortal Thirty-Three." Following the United Kingdom's diplomatic intervention, Brazil and Argentina recognized Uruguay as an independent nation. The country's first constitution was adopted in 1830.

Young soccer players are so enthusiastic about their game that they risk the bruises that can come from playing on hard city streets. Soccer is by far the most popular sport in Uruguay, and many children begin playing soccer as soon as they can walk. The first World Cup championship games were played in Montevideo in 1930. Today, soccer games draw huge crowds to city stadiums. Uruguayans also enjoy basketball and rugby. Gaucho rodeos, called domos, are also popular events.

A ranch hand gathers wool that has just been shorn from a sheep. The wool will soon be shipped to a factory in Montevideo. Textiles and wool are among Uruguay's leading exports.

URUGUAY TODAY

For most of its history as an independent nation, Uruguay has had a democratic government. However, conflict between its two major political parties—the Colorados and the Blancos—has caused much internal political strife.

Most of the members of the Colorado Party come from the cities, while the Blanco Party represents the interests of the rural people and landowners. For many Uruguayans, party membership is a matter of family tradition.

The Colorados and the Blancos

Throughout the mid-1800's, the Colorados and the Blancos fought for control of the government. During this time, foreign governments often interfered in Uruguay's affairs by supporting one of these groups. By 1870, the Colorados had become the dominant political party, mainly because a huge wave of immigration dramatically increased Montevideo's population.

In 1903, José Batlle y Ordóñez, a strong believer in democratic principles and social justice, was elected president of Uruguay. Under his leadership, the ruling Colorado Party passed wide-ranging laws that made Uruguay a model of democracy, social reform, and economic stability.

A new constitution

Economic problems began to develop for Uruguay in the 1950's. Foreign trade decreased, while the cost of social programs increased. In 1951, Uruguay ratified a new constitution that abolished the presidency and established a nine-member National Council of Government.

Although the council allowed the Colorados and the Blancos to share political power, it proved inefficient in dealing with the nation's economic problems. In 1967, another new constitution reestablished a presidential government.

As Uruguay's economic problems worsened, public unrest increased. Terrorist violence erupted, and one of the antigovernment groups, the Tupamaros, kidnapped and murdered Uruguayan and foreign officials. In 1973,

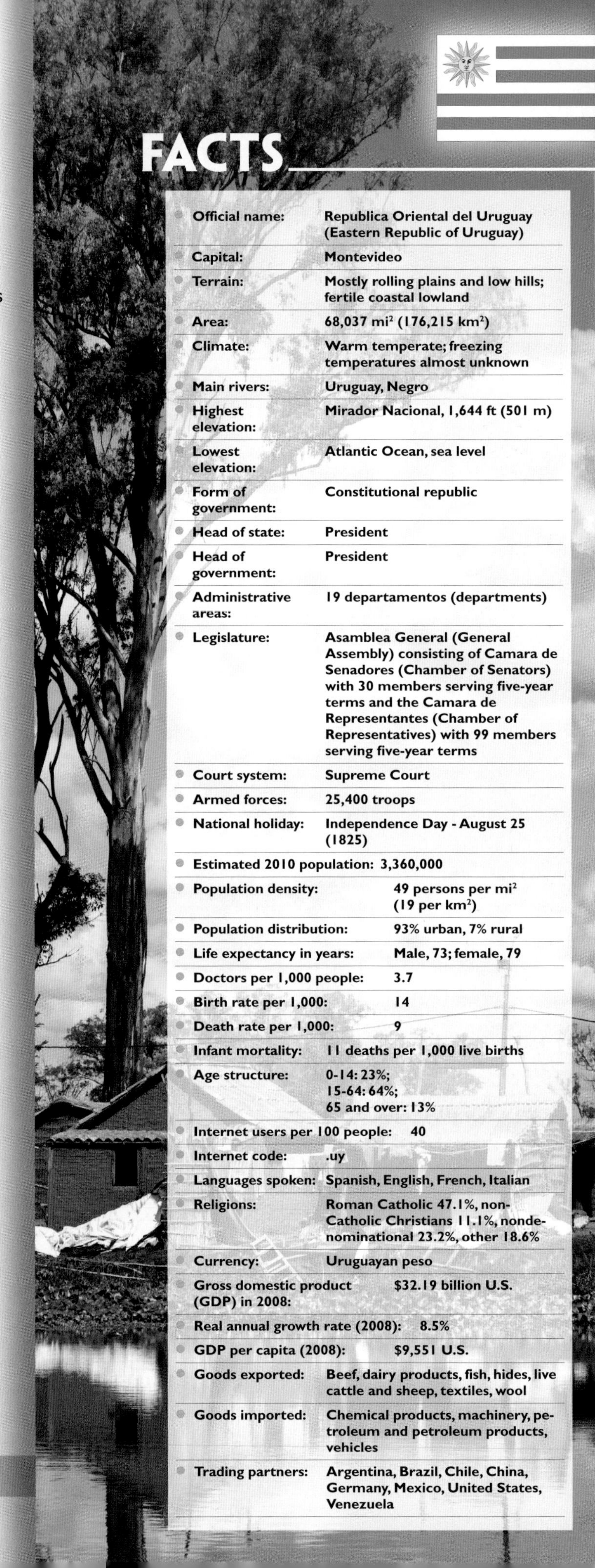

FACTS

Official name:	Republica Oriental del Uruguay (Eastern Republic of Uruguay)
Capital:	Montevideo
Terrain:	Mostly rolling plains and low hills; fertile coastal lowland
Area:	68,037 mi² (176,215 km²)
Climate:	Warm temperate; freezing temperatures almost unknown
Main rivers:	Uruguay, Negro
Highest elevation:	Mirador Nacional, 1,644 ft (501 m)
Lowest elevation:	Atlantic Ocean, sea level
Form of government:	Constitutional republic
Head of state:	President
Head of government:	President
Administrative areas:	19 departamentos (departments)
Legislature:	Asamblea General (General Assembly) consisting of Camara de Senadores (Chamber of Senators) with 30 members serving five-year terms and the Camara de Representantes (Chamber of Representatives) with 99 members serving five-year terms
Court system:	Supreme Court
Armed forces:	25,400 troops
National holiday:	Independence Day - August 25 (1825)
Estimated 2010 population:	3,360,000
Population density:	49 persons per mi² (19 per km²)
Population distribution:	93% urban, 7% rural
Life expectancy in years:	Male, 73; female, 79
Doctors per 1,000 people:	3.7
Birth rate per 1,000:	14
Death rate per 1,000:	9
Infant mortality:	11 deaths per 1,000 live births
Age structure:	0-14: 23%; 15-64: 64%; 65 and over: 13%
Internet users per 100 people:	40
Internet code:	.uy
Languages spoken:	Spanish, English, French, Italian
Religions:	Roman Catholic 47.1%, non-Catholic Christians 11.1%, nondenominational 23.2%, other 18.6%
Currency:	Uruguayan peso
Gross domestic product (GDP) in 2008:	$32.19 billion U.S.
Real annual growth rate (2008):	8.5%
GDP per capita (2008):	$9,551 U.S.
Goods exported:	Beef, dairy products, fish, hides, live cattle and sheep, textiles, wool
Goods imported:	Chemical products, machinery, petroleum and petroleum products, vehicles
Trading partners:	Argentina, Brazil, Chile, China, Germany, Mexico, United States, Venezuela

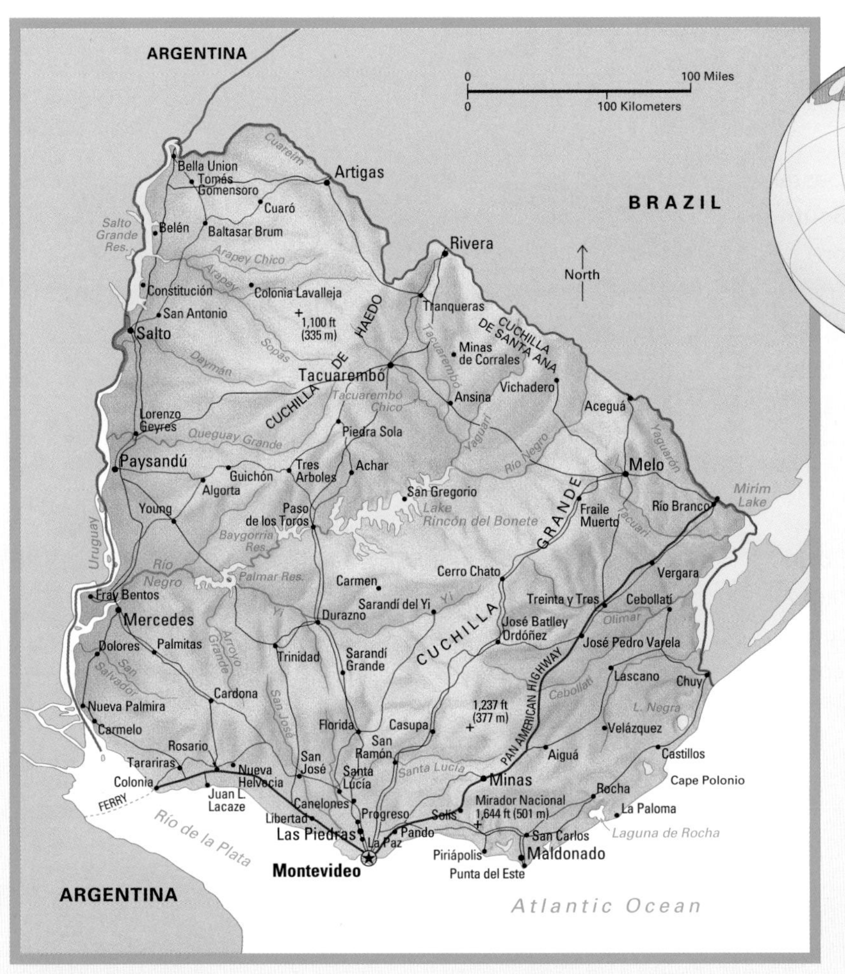

Uruguay lies at the mouth of the Rio de la Plata. Once one of the wealthiest countries in South America, Uruguay now has serious economic problems. Unemployment, along with less money available for social programs, has hurt many Uruguayans.

military officers took over the government and forced President Juan Maria Bordaberry to dissolve the national legislature. Three years later, they removed Bordaberry from office and named Aparicio Mendez to the presidency. General Gregorio Alvarez succeeded Mendez in 1981.

Despite the bitter protests of the Uruguayan people, military rule continued during the early 1980's. Negotiations between military and party leaders led to democratic elections and a return to civilian government in 1984. Julio María Sanguinetti, leader of the Colorado Party, took office as president in 1985.

With the return of democracy to Uruguay, its new leadership had to rebuild the country's economy. Export income dropped sharply during the early and middle 1980's. As a result, the amount of money available for social programs decreased. The country's foreign debt skyrocketed. In 1994, Sanguinetti again was re-elected president. In 1999, Jorge Batlle Ibáñez, also of the Colorado Party, was elected president.

During the late 1990's and early 2000's, Uruguay experienced a severe recession, caused partly by economic troubles in neighboring Argentina and Brazil. In 2004, Tabaré Ramón Vázquez Rosas of the Socialist Party was elected president. Vázquez was the first leftist to be elected president of Uruguay. Vázquez improved Uruguay's economy. Nevertheless, in the 2009 election, voters chose José Mujica, a senator and former Marxist guerrilla, as president.

The popular resort city of Punta del Este lies on the southeastern tip of Uruguay. Its beautiful sandy beaches draw throngs of vacationers from both Uruguay and neighboring Argentina. Uruguay has many magnificent beaches along its Atlantic coast.

LAND AND ECONOMY

Uruguay can be divided into two land regions: the coastal plains and the interior lowlands. Unlike many other South American countries, Uruguay features no dramatic contrasts in its landscape, but it is a lovely, scenic country. Miles of sandy white beaches stretch along its coast, and sparkling rivers flow through the gently rolling grasslands of the interior.

The coastal plains

The coastal plains extend in a narrow arc along the Uruguay River, the Río de la Plata, and the Atlantic Ocean. The Uruguay River forms the country's western border with Argentina. Although the coastal plains cover only about a fifth of Uruguay, most of the nation's population is concentrated in this region, especially along the southern coast.

Workers in a tannery process the skins and hides that form one of Uruguay's most valuable exports. Most working-class people in Uruguay enjoy a comfortable standard of living, with decent housing and good medical care.

An Uruguayan vineyard worker shows off the luscious grape harvest. Grapes are grown mainly on the coastal plains of Uruguay, where the crop produces enough wine to meet the country's domestic needs.

Gauchos work hard at roundup time on a sheep ranch in the interior of Uruguay. Many gauchos still wear at least part of the traditional costume, which includes baggy trousers tucked into boots, a poncho, and a wide-brimmed hat. Most other Uruguayans today dress much as people do in the United States and Canada.

Punta del Este is a popular resort on the southern tip of Uruguay. The city's beaches on the South Atlantic Ocean are a major tourist attraction.

The capital city of Montevideo, located on the southern coast, is also the nation's commercial, political, and intellectual center. About half of Uruguay's nearly 3-1/2 million people live there. Montevideo is a bustling city with many lovely parks, impressive monuments, treelined avenues, and beautiful beaches. Montevideo offers a variety of cultural and recreational opportunities.

Most of the people who live in Montevideo—and throughout Uruguay—are descended from Spanish and Italian immigrants who settled in Uruguay during the 1800's and early 1900's. Other groups include people of English, French, German, and Eastern European descent. Mestizos (people of mixed European and Indian ancestry) make up between 5 and 10 percent of Uruguay's population. Less than 5 percent are descended from African people who were brought to Uruguay to work as slaves. The inhabitants of pure Indian ancestry disappeared almost entirely by the late 1700's.

Many people in Montevideo work in service industries such as banks, health care facilities, schools, transportation, and communications. Others work in factories, or as laborers or household servants. Along the Atlantic coast, many Uruguayans are employed by hotels, restaurants, and resorts that serve the tourist industry.

The western and southwestern coastal plains have Uruguay's richest soil. In this region, family farms and large plantations produce enough barley, corn, potatoes, rice, soybeans, sugar cane, sunflower seeds, and wheat to feed Uruguay's large urban population. The rest of the land is taken up by huge cattle and sheep ranches.

The interior lowlands

The interior lowlands cover most of Uruguay. The grass-covered plains and hills of the interior lowlands make ideal pastureland for livestock. As a result, sprawling cattle and sheep ranches cover the countryside. The production costs of raising livestock are low in this region, and the quality of the product is high.

UZBEKISTAN

Uzbekistan is an independent country that lies in the foothills of the Tian Shan and Pamir mountains and extends to the Aral Sea. It is bordered by Afghanistan in the south, Turkmenistan in the southwest, Kazakhstan in the west and north, and Kyrgyzstan and Tajikistan in the east. The Kyzylkum desert covers a large part of northwest Uzbekistan.

Uzbekistan is a member of the Commonwealth of Independent States (CIS), which was formed in late 1991. Uzbekistan was formerly a republic of the Soviet Union.

History

Two major rivers—the Amu Darya and Syr Darya—flow through Uzbekistan. The fertile valleys of these rivers once served as oases for caravans along the great Silk Road, a major trade route between the East and West opened by Kushan emperors about A.D. 50.

The cities of Tashkent, Samarqand, and Khiva—where the routes between Europe and the Middle East and between China and India crisscrossed—were centers of world trade during ancient times. The colorful markets, where silk, spices, and other luxury goods—as well as ideas and customs—were once traded, remain today.

The area that is now Uzbekistan was the ancient Persian province of Sogdiana, one of the world's oldest civilized regions. During the 300's B.C., it was conquered by Alexander the Great. Turkic nomads took control of the region in the A.D. 500's, and in the 700's, the Arabs arrived and introduced the people to the religion of Islam.

The Seljuk Turks gained control of the region in the 1100's but were driven out by the Mongols in the 1200's. Then, in the 1300's, a descendant of Genghis Khan named Tamerlane made Samarqand the center of his vast empire. In the early 1500's, people from the Golden Horde—a once-powerful region of the Mongol Empire—swept down from the northwest and established the Uzbek Empire.

During the 1600's, the Uzbek Empire broke up into separate states, which were taken over by Russia in the late 1800's. In 1924, the region became the Uzbek Soviet Socialist Republic.

FACTS

Official name:	Uzbekiston Respublikasi (Republic of Uzbekistan)
Capital:	Tashkent
Terrain:	Mostly flat-to-rolling sandy desert with dunes; broad, flat intensely irrigated river valleys in east surrounded by mountainous Tajikistan and Kyrgyzstan; shrinking Aral Sea in west
Area:	172,742 mi² (447,400 km²)
Climate:	Mostly midlatitude desert, long, hot summers, mild winters; semiarid grassland in east
Main rivers:	Amu Darya, Syr Darya, Zeravshan
Highest elevation:	Peak in the Gissar mountain range, 15,233 ft (4,643 m)
Lowest elevation:	Sarykamysh Lake (seasonal salt lake bed), 65 ft (20 m) below sea level
Form of government:	Republic
Head of state:	President
Head of government:	Prime minister
Administrative areas:	12 wiloyatlar (provinces), 1 respublikasi (autonomous republic), 1 shahar (city)
Legislature:	Oliy Majlis (Supreme Assembly) consisting of a Senate with 100 members serving five-year terms and a Legislative Chamber with 120 members serving five-year terms
Court system:	Supreme Court
Armed forces:	67,000 troops
National holiday:	Independence Day - September 1 (1991)
Estimated 2010 population:	28,133,000
Population density:	163 persons per mi² (63 per km²)
Population distribution:	63% rural, 37% urban
Life expectancy in years:	Male, 66; female, 73
Doctors per 1,000 people:	2.7
Birth rate per 1,000:	21
Death rate per 1,000:	5
Infant mortality:	36 deaths per 1,000 live births
Age structure:	0-14: 32%; 15-64: 63%; 65 and over: 5%
Internet users per 100 people:	9
Internet code:	.uz
Languages spoken:	Uzbek (official), Russian, Tajik
Religions:	Muslim 88% (mostly Sunnis), Eastern Orthodox 9%, other 3%
Currency:	Uzbek som
Gross domestic product (GDP) in 2008:	$27.92 billion U.S.
Real annual growth rate (2008):	8.9%
GDP per capita (2008):	$1,001 U.S.
Goods exported:	Automobiles, copper and other metals, cotton, energy products
Goods imported:	Food, iron and steel, machinery, transportation equipment
Trading partners:	China, Kazakhstan, Russia, Turkey, Ukraine

A marketplace in Buxoro is reminiscent of the ancient bazaars that once stood on this site. One of the largest cities in Uzbekistan today, Bukhara was once a stopping place for caravans along the ancient trade route known as the Silk Road.

Uzbekistan remained under the strict control of the Soviet Union until1991. In the midst of political upheaval in the Soviet Union in August 1991, Uzbekistan declared its independence. It joined the Commonwealth of Independent States, a loose confederation of former Soviet republics.

In December 1991, Islam Karimov was elected president. Karimov won a second term in 2000 and a third term in 2007. His government harassed opposition leaders, banning some political groups and all religious parties. Strife between different ethnic groups broke out from time to time, and the country has refused to grant dual citizenship to its Russian minority.

Land and people

Although most of Uzbekistan consists of plains and deserts, the region is rich in mineral resources, including coal, copper, gold, natural gas, and petroleum. Oil fields stand in the Fergana Valley, a populous area along the upper part of the Syr Darya, and western Uzbekistan has large natural gas deposits. Hydroelectric power stations along the rivers supply energy for the large-scale irrigation systems used in agricultural production.

Cotton is the chief farm product. Farmers also raise livestock. Other important products include grapes, melons and other fruits, milk, rice, vegetables, and wool from the karakul, a breed of sheep raised in the country.

The majority of the people who live in Uzbekistan are Uzbeks and follow the religion of Islam. The remaining population consists mainly of Russian people. Today, Uzbekistan is a blend of ancient Asian and modern European cultures.

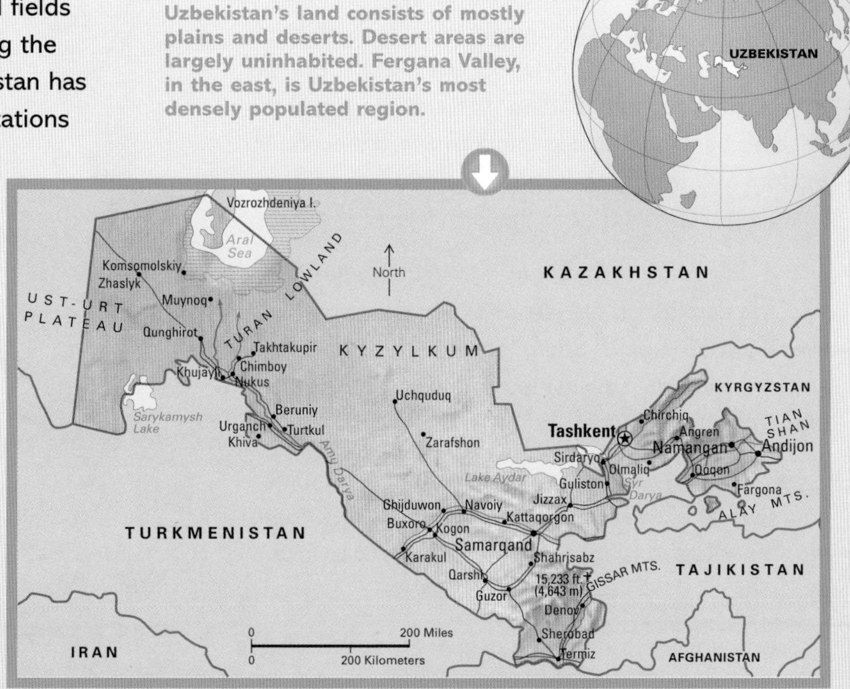

Uzbekistan's land consists of mostly plains and deserts. Desert areas are largely uninhabited. Fergana Valley, in the east, is Uzbekistan's most densely populated region.

VANUATU

Eighty islands in the southwest Pacific Ocean make up the country of Vanuatu. The islands form a Y-shaped chain that extends about 500 miles (800 kilometers) from north to south. Most of the islands have narrow coastal plains and mountainous interiors, and several have active volcanoes. Vanuatu has a tropical wet climate—always hot and wet, with heavy precipitation throughout the year. Many houses, in fact, have thatched roofs to keep rain from dripping in. Savanna grassland and bush are prevalent in the southern islands, while tropical rain forests cover much of the northern islands.

Way of life

More than 90 percent of Vanuatu's estimated 236,00 people are Melanesians. Asians, Europeans, and Polynesians make up the rest of the population. Most of the people live in rural villages. Many village houses are made of wood, bamboo, and palm leaves.

More than 100 languages are spoken in Vanuatu. Bislama, a type of Pidgin English that combines English words and Melanesian grammar, is widely used.

Agriculture is the country's chief economic activity. Rural families produce nearly all their own food, and some families produce copra (dried coconut meat) for sale. They grow fruits and vegetables, raise chickens and hogs, and catch fish. Beef, cacao, and copra are among the country's leading exports. Tourism and offshore banking are also important to the economy. Visitors and residents alike enjoy boating on the sparkling blue water near the islands' sandy shores.

The only urban communities in Vanuatu are Port-Vila, the capital, on the island of Efate, and Santo, on the island of Espiritu Santo. There are few good roads and no railroads. Small ships and airplanes are the primary means of transportation in the islands.

Vanuatu has hundreds of elementary schools and several high schools. A majority of the people are Christians, and most of the rest practice local religions.

Vanuatu is a republic. The country's laws are made by a Parliament, whose members are elected by the

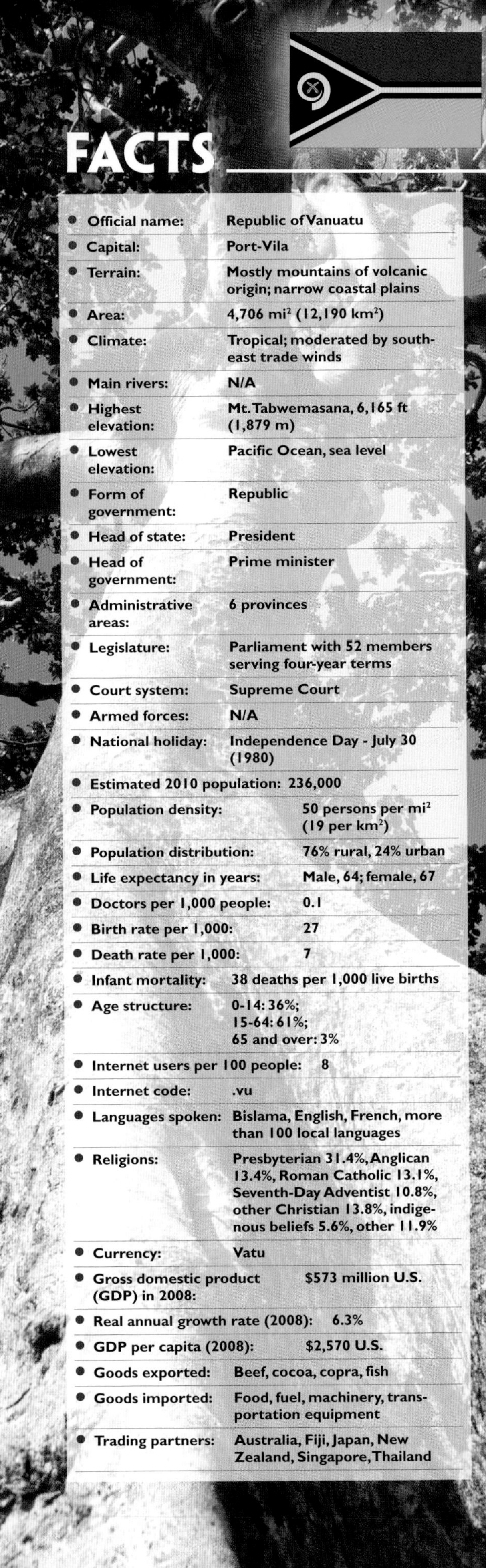

FACTS

Official name:	Republic of Vanuatu
Capital:	Port-Vila
Terrain:	Mostly mountains of volcanic origin; narrow coastal plains
Area:	4,706 mi^2 (12,190 km^2)
Climate:	Tropical; moderated by southeast trade winds
Main rivers:	N/A
Highest elevation:	Mt. Tabwemasana, 6,165 ft (1,879 m)
Lowest elevation:	Pacific Ocean, sea level
Form of government:	Republic
Head of state:	President
Head of government:	Prime minister
Administrative areas:	6 provinces
Legislature:	Parliament with 52 members serving four-year terms
Court system:	Supreme Court
Armed forces:	N/A
National holiday:	Independence Day - July 30 (1980)
Estimated 2010 population:	236,000
Population density:	50 persons per mi^2 (19 per km^2)
Population distribution:	76% rural, 24% urban
Life expectancy in years:	Male, 64; female, 67
Doctors per 1,000 people:	0.1
Birth rate per 1,000:	27
Death rate per 1,000:	7
Infant mortality:	38 deaths per 1,000 live births
Age structure:	0-14: 36%; 15-64: 61%; 65 and over: 3%
Internet users per 100 people:	8
Internet code:	.vu
Languages spoken:	Bislama, English, French, more than 100 local languages
Religions:	Presbyterian 31.4%, Anglican 13.4%, Roman Catholic 13.1%, Seventh-Day Adventist 10.8%, other Christian 13.8%, indigenous beliefs 5.6%, other 11.9%
Currency:	Vatu
Gross domestic product (GDP) in 2008:	$573 million U.S.
Real annual growth rate (2008):	6.3%
GDP per capita (2008):	$2,570 U.S.
Goods exported:	Beef, cocoa, copra, fish
Goods imported:	Food, fuel, machinery, transportation equipment
Trading partners:	Australia, Fiji, Japan, New Zealand, Singapore, Thailand

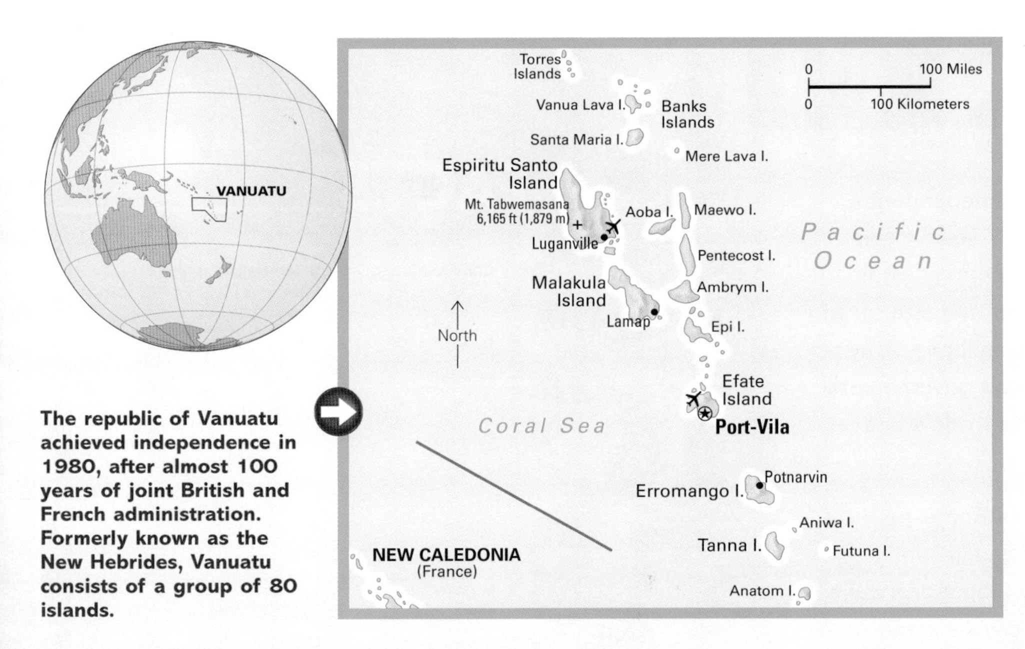

The republic of Vanuatu achieved independence in 1980, after almost 100 years of joint British and French administration. Formerly known as the New Hebrides, Vanuatu consists of a group of 80 islands.

people to four-year terms. A prime minister, who heads the majority party in Parliament, runs the government with the aid of a Council of Ministers. The Parliament and regional council presidents elect a president, whose role is chiefly ceremonial. The government publishes a newspaper and operates a radio station.

History

Melanesians have lived in what is now Vanuatu for at least 3,000 years. In 1606, the commander of a Spanish expedition from Peru became the first European to see the islands. The British explorer James Cook mapped the region in 1774 and named the islands the *New Hebrides* after the Hebrides Islands of Scotland.

British and French traders, missionaries, and settlers began coming to the islands during the 1820's. Settlers established vast plantations on the fertile land and grew coconut palms, coffee, and cocoa. They also set up large beef cattle farms.

In 1887, the United Kingdom and France set up a joint naval commission to oversee the area. In 1906, the commission was replaced by a joint British and French government called a *condominium*. British and French interests clashed, however. Each nation had its own colonial administrator and its own police force on the islands, and each insisted on using its own currency and language.

During World War II (1939–1945), the New Hebrides became an important Allied military base. U.S. troops built many roads, bridges, and airstrips there. In 1945, foreigners owned more than a third of the land area. But during the 1960's, islanders began a movement for independence, and the New Hebrides became the independent nation of Vanuatu on July 30, 1980.

In 1987, a cyclone struck the islands, causing much damage and loss of life.

The peaceful town of Port-Vila is Vanuatu's capital. A fierce cyclone struck in 1987, leaving behind severe damage.

VATICAN CITY

The State of Vatican City is the smallest independent country in the world. It occupies only 0.17 square miles (0.44 square kilometers) on Vatican Hill, entirely within the city of Rome. Although it is only about the size of an average city park, Vatican City influences millions of people all over the world. It is the spiritual and governmental center of the Roman Catholic Church, the world's largest Christian church.

Vatican City is situated in northwestern Rome, just west of the Tiber River. Most of the Vatican is surrounded by high stone walls. Within these walls stand a number of lovely buildings, courtyards, landscaped gardens, and quiet streets. Most of the area is taken up by St. Peter's Church.

Pope Benedict XVI, a German, was elected in 2005. The pope is head of the Roman Catholic Church and bishop of Rome, as well as head of the Vatican's legislative, executive, and judicial affairs.

Origins of the Vatican

According to tradition, Saint Peter was crucified on Vatican Hill and buried nearby. The early popes erected Vatican City on the site of a shrine thought to have marked Saint Peter's tomb.

In the A.D. 300's, the Christian emperor Constantine the Great built a *basilica* (a church with certain ceremonial privileges) where the shrine stood. Gradually, the Vatican Palace and other buildings were constructed around the basilica. Beginning in the 1500's, St. Peter's Church was built on the site of the Old Basilica of Constantine.

During the Middle Ages, the popes gained control over much of central Italy, an area once known as the Papal States. But with the unification of Italy in 1861, the Kingdom of Italy took over the Papal States. Rome became the capital of Italy in 1871. In protest, Pope Pius IX and his successors withdrew inside the Vatican and refused to deal with the Italian government.

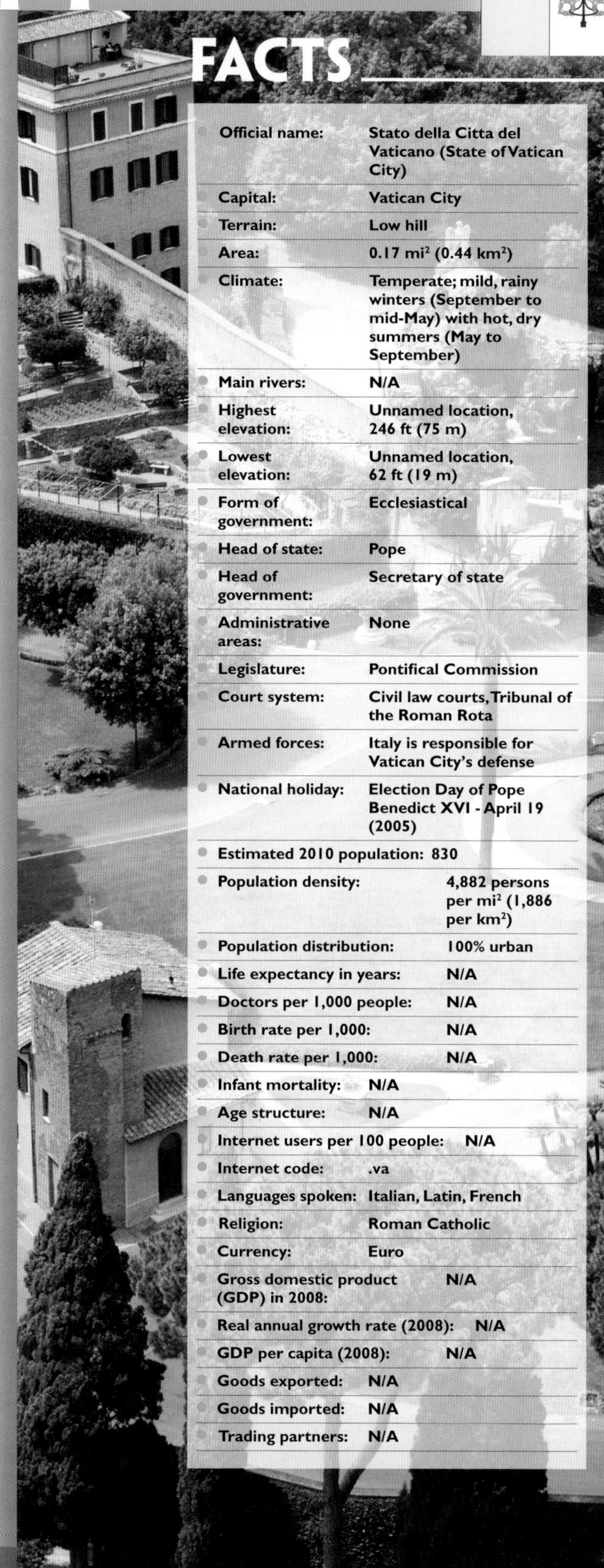

FACTS

Official name:	Stato della Citta del Vaticano (State of Vatican City)
Capital:	Vatican City
Terrain:	Low hill
Area:	0.17 mi² (0.44 km²)
Climate:	Temperate; mild, rainy winters (September to mid-May) with hot, dry summers (May to September)
Main rivers:	N/A
Highest elevation:	Unnamed location, 246 ft (75 m)
Lowest elevation:	Unnamed location, 62 ft (19 m)
Form of government:	Ecclesiastical
Head of state:	Pope
Head of government:	Secretary of state
Administrative areas:	None
Legislature:	Pontifical Commission
Court system:	Civil law courts, Tribunal of the Roman Rota
Armed forces:	Italy is responsible for Vatican City's defense
National holiday:	Election Day of Pope Benedict XVI - April 19 (2005)
Estimated 2010 population:	830
Population density:	4,882 persons per mi² (1,886 per km²)
Population distribution:	100% urban
Life expectancy in years:	N/A
Doctors per 1,000 people:	N/A
Birth rate per 1,000:	N/A
Death rate per 1,000:	N/A
Infant mortality:	N/A
Age structure:	N/A
Internet users per 100 people:	N/A
Internet code:	.va
Languages spoken:	Italian, Latin, French
Religion:	Roman Catholic
Currency:	Euro
Gross domestic product (GDP) in 2008:	N/A
Real annual growth rate (2008):	N/A
GDP per capita (2008):	N/A
Goods exported:	N/A
Goods imported:	N/A
Trading partners:	N/A

The pope delivers an Easter message from a balcony above St. Peter's Square. St. Peter's Square was completed in 1667 and contains an obelisk brought from Egypt about A.D. 37.

It was not until 1929 that the Treaty of the Lateran resolved the status of Vatican City. By this treaty, the pope gave up all claim to the Papal States, and Italy recognized the independence of the State of Vatican City.

The buildings of Vatican City

Vatican City includes a number of important buildings. The Vatican Palace, with well over 1,000 rooms, is a group of connected buildings clustered around several open courts. The palace—which includes various chapels, apartments, museums, and other rooms—is also the residence of the pope.

Among the palace buildings are the Vatican Museums—a vast collection of priceless art treasures. The many rooms and chapels of the museums are decorated with masterpieces created by history's greatest artists, including Fra Angelico, Pinturiccio, Raphael, Titian, and Leonardo da Vinci.

In the famous Sistine Chapel, Michelangelo's ceiling frescoes tell the Biblical stories of the creation of the world, the fall of humanity, and the Flood. The famous statues *Apollo Belvedere* and the *Laocoön* are part of the museums' collection of ancient sculpture. A controversial restoration of the frescoes ended in April 1994, when *The Last Judgment* was unveiled. Critics of the project said the cleaning destroyed subtleties of Michelangelo's shading, but most scholars approved of the work.

Administering the Vatican

The population of Vatican City—about 830—consists mainly of nuns and priests who work in administrative positions in the Roman Catholic Church. The pope is the absolute ruler of Vatican City, but the day-to-day affairs are the responsibility of the Pontifical Commission for the State of Vatican City, whose members are appointed by the pope. A governor, whose duties resemble those of a mayor, directs the city's administration.

Cardinals and bishops gather in the Sistine Chapel during the bishop's synod, an assembly called to discuss church affairs. Cardinals from all over the world elect the pope.

Vatican City, an independent state since the Treaty of Lateran was signed in 1929, lies entirely within the city of Rome. In addition to a number of important church buildings, Vatican City includes large blocks of apartment buildings for its citizens—mostly the nuns and priests who work there. Gardens cover the western half of the Vatican, while the famous Sistine Chapel and St. Peter's Church lie in the southeastern corner.

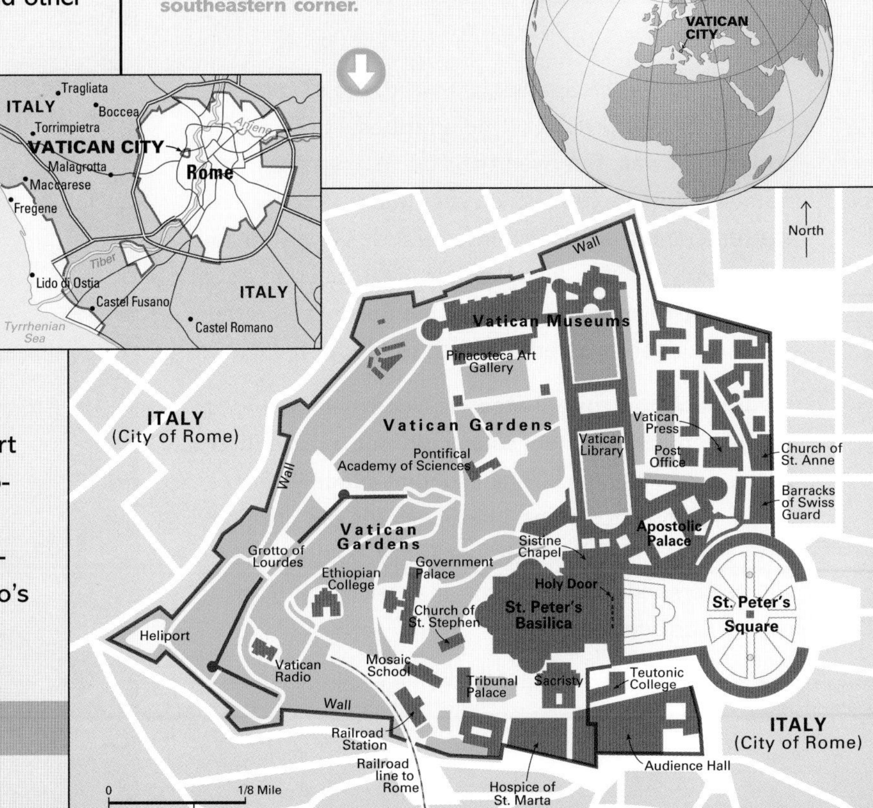

VENEZUELA

Venezuela, a land of sprawling, modern cities, spectacular scenery, and gushing oil wells, lies on the north coast of South America. Once one of the poorer nations of South America, with an agriculture-based economy, Venezuela has been transformed by its petroleum industry. Now it is one of the continent's wealthiest and most rapidly changing countries.

Although the face of Venezuela is changing in many ways, reminders of this nation's long history can still be seen. The remarkable story of this South American country is written in the faces of its people—from the Indian tribes in the southern forests to the residents of Spanish colonial buildings in the northern city of Coro and those in the high-rise apartment buildings of the capital city of Caracas.

Spanish conquest and colonial rule

Christopher Columbus was the first European to reach Venezuela. He landed on Venezuela's Paria Peninsula in 1498, during his third voyage to the New World. Later, in the northwest region of the country, Spanish explorers discovered Indian houses built on stilts over the waters of Lake Maracaibo and the connecting gulf that opens northward to the Caribbean Sea. These houses reminded the Spaniards of the villas lining the canals of Venice, so the explorers named the region *Venezuela*, which is Spanish for *Little Venice*.

Long before the Spaniards arrived, the land was inhabited by two groups of Indian tribes—the Carib and the Arawak. But even their bravest warriors were no match for the ruthless *conquistadores* (conquerors), and the Spanish invaders took over the land in the early 1500's. Many Indians were killed in battle or died of diseases brought by the Europeans. Others starved or were worked to death as laborers.

Although the coastal areas of Venezuela provided pearls and salt, the new colony did not prosper under the Spanish. Its agricultural economy made it one of Spain's poorest colonies. The Spaniards did not find in Venezuela the incredible hordes of gold and silver which Hernando Cortés found in Mexico and Francisco Pizarro found in Peru.

Struggle for independence

In 1730, Spain gave the Royal Guipuzcoana Company of Caracas all rights to trade in Venezuela. The colonists, who bitterly resented the company's activities, began to long for independence from Spain.

The colony declared its independence on July 5, 1811, but Spanish forces continued to occupy much of the country. In 1819, Simón Bolívar, one of the chief leaders of the independence movement, set up a republic called Gran Colombia, which eventually included what are now Venezuela, Colombia, Ecuador, and Panama. Not until 1821 did Venezuela gain true independence from Spain. In that year, Bolívar defeated the Spanish at Carabobo, near Valencia.

By 1831, Venezuela had broken away from the republic of Gran Colombia and elected General José Antonio Páez as the nation's first president.

A series of civil wars broke out in 1846 and lasted until 1870, when the dictatorial caudillo (leader) Antonio Guzmán Blanco seized power. Guzmán Blanco brought order to Venezuela and encouraged investment in the development of the country, but he was overthrown in 1888. The nation was then ruled by a series of other caudillos.

Although reformist political parties were founded in Venezuela in the mid-1930's, democracy was slow to come. In 1947, the people elected Rómulo Gallegos of the Acción Democrática (AD) party as president, but Gallegos was overthrown by the army in 1948. Since 1958, every president of Venezuela has come from either AD or the nation's other major political party—the Comité de Organización Política Electoral Independiente (COPEI), also known as the Social Christian Party.

Today, Venezuela's primary struggle is to lessen its dependence on petroleum and to minimize the economic instability caused by price changes in this important resource.

VENEZUELA TODAY

Once valued for little more than the salt mines on the Araya Peninsula and pearls along the north coast, Venezuela has become one of South America's wealthiest and most progressive nations. But the wealth is not distributed evenly among the people, and poverty is a major problem in some areas. Another difficulty is the economic instability created by changes in the price of petroleum.

Economic and political developments

Since the 1920's, when the nation's oil industry began to boom, Venezuela's leaders have used the income from petroleum exports to pay for massive industrial and modernization programs. The skyline of Caracas, the capital and largest city, reflects the changes funded by the petroleum profits. High-rise buildings, wide boulevards, and multilaned expressways have transformed Caracas. What was once a quiet colonial city is now a bustling and crowded metropolis.

By the 1980's, profits from the oil industry had created a well-to-do middle class. However, the steel and glass towers of Caracas could not hide the dreadful poverty of the city's slums. Beginning in the 1960's, the government tried to improve the conditions of the poor through public housing and other programs. But government services had difficulty keeping up with the swelling population, and many unskilled laborers continued to crowd into squatter settlements along the outskirts of Caracas and other Venezuelan cities.

Rising prices led to riots in 1989. Three years later, an attempted coup against President Carlos Andrés Pérez failed. The coup leader, Colonel Hugo Chávez Frías, was jailed but pardoned in 1994. President Pérez was removed from office in 1993 on charges of embezzling public funds.

In January 1994, Venezuela's second largest bank failed, causing a financial panic. In February, Rafael Caldera Rodríguez was elected president, a post he had held from 1969 to 1974. Soon after Caldera took office, however, the country suffered a banking crisis. The government then took steps to tighten control over the country's banking system.

In 1998, Hugo Chávez was elected president. In a 1999 referendum, Venezuelans approved a new constitution that allowed the president to be reelected for a second term.

FACTS

Official name:	Republica Bolivariana de Venezuela (Bolivarian Republic of Venezuela)
Capital:	Caracas
Terrain:	Andes Mountains and Maracaibo Basin in northwest; central plains (llanos); Guiana Highlands in southeast
Area:	352,145 mi² (912,050 km²)
Climate:	Tropical; hot, humid; more moderate in highlands
Main rivers:	Orinoco
Highest elevation:	Pico Bolívar, 16,411 ft (5,002 m)
Lowest elevation:	Caribbean Sea, sea level
Form of government:	Federal republic
Head of state:	President
Head of government:	President
Administrative areas:	23 estados (states), 1 distrito federal (federal district), 1 dependencia federal (federal dependency)
Legislature:	Asamblea Nacional (National Assembly) with 167 members serving five-year terms
Court system:	Tribuna Suprema de Justicia (Supreme Court of Justice)
Armed forces:	115,000 troops
National holiday:	Independence Day - July 5 (1811)
Estimated 2010 population:	28,920,000
Population density:	82 persons per mi² (32 per km²)
Population distribution:	94% urban, 6% rural
Life expectancy in years:	Male, 70; female, 76
Doctors per 1,000 people:	1.9
Birth rate per 1,000:	22
Death rate per 1,000:	5
Infant mortality:	17 deaths per 1,000 live births
Age structure:	0-14: 31%; 15-64: 64%; 65 and over: 5%
Internet users per 100 people:	25
Internet code:	.ve
Languages spoken:	Spanish (official), numerous indigenous dialects
Religions:	Roman Catholic 96%, Protestant 2%, other 2%
Currency:	Bolívar fuerte
Gross domestic product (GDP) in 2008:	$319.44 billion U.S.
Real annual growth rate (2008):	4.9%
GDP per capita (2008):	$11,363 U.S.
Goods exported:	Mostly: petroleum Also: aluminum, chemicals, iron and steel
Goods imported:	Chemicals, food and live animals, machinery, transportation equipment
Trading partners:	Brazil, China, Colombia, Germany, Japan, Mexico, United States

Venezuela lies on the north coast of South America, along the Caribbean Sea. Mountain ranges extend across much of northern Venezuela, which is the most densely populated region of the country. Caracas, the capital and largest city, lies in this region. Vast plains called the Llanos spread across central Venezuela. High plateaus and low mountains cover the south.

Chávez was reelected in 2000 and again in 2006. His attempts to increase his control over Venezuela's state-run oil company led business, labor, and political opposition leaders to organize protests. Nevertheless, a referendum in 2009 eliminated term limits for the president and other elected government officials, and in 2010, the outgoing National Assembly gave Chávez the power to rule by decree for 18 months. Some critics claimed Chávez planned to use the decree powers to overrule the newly elected legislature.

Social progress

Compared with some other Latin American countries, which have a strict class system based on ancestry, Venezuelans are not as rigidly segregated on the basis of racial or class differences. However, poverty remains a widespread problem.

Also, beginning in the 1990's, the standard of living of Venezuela's middle class—made up of business people, government workers, and doctors, lawyers, teachers, and other professionals—has fallen. Economists blame the decline on cyclic changes in the price of oil. Government programs since the 1960's have established a social security system and improved public health and education. Children between the ages of 7 and 13 must attend school, and free university education is available. Although such programs have helped the poor, they have not stopped the decline of the middle class. Unemployment is a major problem for both groups.

A Venezuelan woman stands in the doorway of a brightly painted house in Maracaibo. In many of Venezuela's cities, these traditional, Spanish-style houses are being replaced by high-rise apartment buildings.

LAND AND ECONOMY

Venezuela lies on the north coast of South America, along the Caribbean Sea. In addition to its mainland territory, Venezuela has 72 islands off its coast. Venezuela also claims a large part of Guyana, its neighbor to the east. The Guyanese territory in question includes more than 50,000 square miles (130,000 square kilometers) of land west of the Essequibo River.

Landscape

A nation of extraordinary natural beauty, Venezuela boasts spectacular waterfalls, swift-flowing rivers, soaring snow-capped mountain peaks, and broad, fertile valleys. Its four land regions reflect the country's scenic variety.

The Maracaibo Basin is located in northwestern Venezuela. It consists of Lake Maracaibo—the largest lake in South America—and the lowlands surrounding it.

The Andean Highlands form a natural boundary between the Maracaibo Basin and the gently sloping plains of the *Llanos*. Most of Venezuela's people live in the Andean Highlands, which include the Mérida Range, the Central Highlands, and the Northeastern Highlands. The capital city of Caracas lies close to the north-central coast, where the Andean Highlands meet the Caribbean Sea.

A farmer leads his horse to pasture in a lush valley nestled in the Andean Highlands. The two parallel ranges of the Central Highlands enclose many such valleys, where small farms thrive.

Huge oil tankers lie at the loading wharfs in the busy port of Barcelona, which serves the oil fields of the eastern Llanos. Petroleum is Venezuela's leading export, and Venezuela is one of the world's largest petroleum exporters.

Huge, sprawling cattle ranches dot the open prairies of the Llanos, where cowhands known as *llaneros* herd their livestock. The region also has some farmland, but the dry climate makes irrigation necessary for such crops as rice and sesame.

The Guiana Highlands, which rise south of the Llanos, cover nearly half of Venezuela and are almost empty of people. Tropical rain forests blanket much of the southern part of the region. The Orinoco River, Venezuela's major waterway, rises in the Guiana Highlands near the Brazilian border.

An abundance of wildlife

Wildlife is plentiful in Venezuela. A variety of animals and plants—including various species of eagles and Venezuela's national flower, the orchid—flourish in the country. Venezuela's magnificent forests are home to such exotic creatures as tapirs, sloths, anteaters, and a variety of monkeys, as well as such colorful tropical birds as caciques, parrots, and macaws. Puma, deer, and vampire bats live high in the mountains, while crocodiles bask in the lowland rivers.

Angel Falls, one of the world's highest waterfalls, creates a magnificent spectacle as the waters plunge 3,212 feet (979 meters) into the Churún River. The waterfall is named for the American pilot Jimmy Angel, who flew over the falls in 1935.

Cacao, bananas, coconuts, mangoes, and palm and rubber trees are grown in the humid tropic zones. In dry areas, cacti and prickly pears grow. Oranges, lemons, avocados, peaches, apricots, tobacco, sugar cane, cotton, rice, coffee, maize, beans, and potatoes are grown in semitropical and temperate areas of the country.

Economy

Venezuela, one of the world's leading oil producers, gets most of its export earnings from oil. However, in an effort to reduce its dependency on oil income, the government has encouraged the growth of manufacturing. The Guri Dam hydroelectric project spearheaded the construction of one of the largest industrial zones in all of Latin America, at Ciudad Guayana. Today, in addition to petroleum-refining plants in Maracaibo, aluminum and steel are produced in Ciudad Guayana. Venezuelan factories also produce cement, processed foods, and textiles.

Although rich in deposits of petroleum, natural gas, bauxite, coal, diamonds, gold, iron ore, and phosphate rock, Venezuela has few areas of very fertile soil. Only about 10 percent of Venezuela's workers are farmers. They grow bananas, coffee, corn, rice, and sorghum. In addition, farmers raise beef and dairy cattle, hogs, and poultry. The most highly developed agricultural region is the basin of Lake Valencia.

Venezuela's abundant wildlife includes:

1. **capuchin**
2. **red howler monkey**
3. **sloth**
4. **tamandua**
5. **oilbird, or guacharo**
6. **black spider monkey**
7. **ocelot**
8. **scarlet ibis**
9. **anaconda**
10. **puma**
11. **skunk**
12. **a reptile called the cayman that is related to the alligator**
13. **boat-billed heron**
14. **jaguar**
15. **anteater**
16. **bushmaster, or rattlesnake**
17. **salamander**

PEOPLE

The people of modern Venezuela are the descendants of three very different ethnic groups: the native Indians, the Spanish colonists, and Africans brought in to work as slaves during colonial times. Almost all Venezuelans speak Spanish, the country's official language, and follow the Roman Catholic religion. Some Indians in remote regions have preserved their tribal languages and carry on their traditional religious practices.

Today, a majority of Venezuela's population is of mixed ancestry, and the remainder are people of unmixed European, African, or indigenous (native) ancestry. The native Indian population of Venezuela was almost completely wiped out during the Spanish conquest of the 1500's. Thousands died in battle, and many others starved to death or fell victim to European diseases. Most of the survivors intermarried with the colonists. As a result, people of unmixed Indian ancestry make up only about 1 percent of the population.

A Venezuelan mestizo woman has mixed white and Indian ancestry. About two-thirds of Venezuela's people are of mixed ancestry.

Since the end of World War II (1939–1945), many Europeans and Colombians have moved to Venezuela. Europeans—mainly from Italy, Spain, and Portugal—were attracted to the country's booming, petroleum-based economy. Refugees from Colombia have increased Venezuela's already swelling population since the 1970's.

In December 1999, heavy rains struck northern Venezuela. Floods and mudslides from the rains killed an estimated 30,000 people.

Plaza Bolívar is the main square in Caracas. The plaza is not only the center of civic, political, commercial, social and tourist activity in the city, but it stands on the grounds where Spanish conqueror Diego de Losada founded Caracas in 1567.

City life

Most of Venezuela's people live in cities and towns. In the country's large urban areas, located mainly in the northern valleys and basins of the Andes near the Caribbean, many people live in terrible

Indians living along Venezuela's rivers travel from village to village in small canoes. Houses on stilts, built by Native Americans of the 1500's on Lake Maracaibo, were the inspiration for the name Venezuela, which is Spanish for Little Venice.

poverty. About half of the people of Caracas live in the *barrios* (slum neighborhoods) on the outskirts of the city. Their tumbledown shacks, called ranchos, stand in stark contrast to the modern skyscrapers and high-rise apartment buildings that dominate the city's skyline.

The Venezuelans are a sociable people who find much to enjoy in a variety of activities. Colorful fiestas brighten the streets at Easter and Christmas. Baseball and soccer games attract multitudes of fans to the city stadiums. And Venezuelans love music and dancing—from the rhythmic *salsa* dance to the exciting, foot-stamping *joropo,* the national folk dance.

The forest Indians

Far from the sunny sidewalk cafés and crowded streets of Caracas, a small group of Venezuelans live much as their ancestors lived thousands of years ago. They are the nomadic tribes of the Guiana Highlands—wanderers who roam the forests and jungles in small bands. They hunt game, fish for turtles, and gather fruits, nuts, and berries. These Indians are one of the few groups of Venezuelans who have resisted modern life and rejected contact with the outside world.

In the far northwest corner of Venezuela, on the land stretching from the western shore of Lake Maracaibo to the green slopes of the Perija Mountains, stand the small villages of the Motilones. The Motilones retreated into the swamps and tropical forests during the Spanish conquest. For 450 years, they kept outsiders at bay with 6-foot (1.8-meter) arrows shot from remarkable black palm bows, which they held between their toes while lying on their backs.

Like the nomadic tribes, the Motilones live much as their ancestors did, cultivating a little land and hunting and fishing for most of their food. As many as 70 people may live in one huge oval hut, its wooden frame held together by rattan cord.

Vietnam, a tropical country on the eastern coast of the Southeast Asian peninsula, is bounded by China on the north, Laos and Cambodia on the west, and the South China Sea on the east. Hanoi is the capital of Vietnam, and Ho Chi Minh City is its largest city.

More than 85 percent of the country's population are Kinh, or ethnic Vietnamese. Other ethnic groups living in Vietnam include the Hmong, the Khmer, the Muong, and the Nung. A number of ethnic Chinese people, known as the Hoa, live mainly in the cities.

Most Vietnamese live in villages on the fertile lands of the coastal plain and on deltas formed by rivers. Many make their living as farmers, raising rice and a few other crops. Some Vietnamese fish for a living, particularly those who live near the coast.

Thousands of years ago, people moved from China to the north and from islands to the south and settled in what is now northern Vietnam. The Vietnamese people, who probably developed out of these groups, date their history from around 200 B.C., when the Chinese general Zhao Tuo established the independent kingdom of Nam Viet.

Early history

The kingdom of Nam Viet, which stretched from central Vietnam into parts of southeastern China, was conquered by China in 111 B.C. and renamed Jiao Zhi. Then, in A.D. 679, the Chinese changed the name to *Annam*, meaning *pacified south*.

China controlled the northern part of the kingdom, but by the A.D. 100's, the kingdoms of Funan and Champa had developed in what is now southern and south-central Vietnam. Funan was conquered by Khmer people during the 500's and 600's, but Champa remained independent until the late 1400's.

In 939, the Chinese withdrew from Annam, and the Vietnamese established an independent state called Dai Co Viet. It endured as an empire for over 900 years. The Ly family ruled Dai Co Viet from 1009 until 1225.

In 1225, the Tran family seized power from the Ly rulers and governed the country until 1400. China invaded Dai Co Viet in 1407, but the Vietnamese drove them out in 1427. The Le family, which had led the fight against China, came to power and renamed the country *Dai Viet* (Great Viet).

VIETNAM

Le rulers held the throne until 1787, when they were removed by the Tay Son brothers. The Tay Son also overthrew two other powerful families, the Nguyen and the Trinh. But Nguyen Anh, a member of the Nguyen family, defeated the Tay Son in 1802. Nguyen Anh then declared himself Emperor Gia Long of all Dai Viet, which he renamed Vietnam.

French rule

Roman Catholic missionaries from France began to arrive in Dai Viet in the 1600's. However, Dai Viet's rulers became suspicious of the missionaries and persecuted them through the early 1800's.

In 1858, French military forces began to attack parts of southern Vietnam, partly to stop the persecution of the missionaries and partly because France wanted to become a colonial power in Vietnam. By 1883, the French forced the Nguyen ruler to sign a treaty that gave France control of all Vietnam. The country was divided into three areas—*Cochin China* (southern Vietnam), *Annam* (central Vietnam), and *Tonkin* (northern Vietnam)—as part of French Indochina.

During World War II (1939–1945), Vietnam was under Japanese control. But after Japan's defeat in August 1945, no single group held power in Vietnam. At that time, Ho Chi Minh, a Vietnamese Communist leader, returned to Vietnam from China as head of the Revolutionary League for the Independence of Vietnam, commonly called the Vietminh. The Vietminh took control of much of northern Vietnam.

In September 1945, Ho proclaimed himself head of the independent Democratic Republic of Vietnam (DRV). The French reestablished control of Cochin China, but they were unable to put down all resistance. On Dec. 19, 1946, fighting broke out between France and the Vietminh—the first battles in a series of wars that would go on almost continuously for nearly 30 years.

Vietnam became divided into two zones—Communist North Vietnam and non-Communist South Vietnam. In the Vietnam War (1957–1975), North Vietnamese and Communist-trained South Vietnamese rebels sought to overthrow the government of South Vietnam and to eventually reunite the country. The United States and the South Vietnamese army tried to stop them, but failed.

VIETNAM TODAY

In April 1975, the Communists gained control of South Vietnam, and in 1976, Vietnam was reunified under Communist rule. The new government initiated a number of programs designed to rebuild the war-torn country. To relieve urban overcrowding and unemployment, large numbers of people were moved from cities to rural areas. The government also sent many southerners to "reeducation" camps, which were essentially concentration camps for political prisoners.

Widespread opposition to these programs led about a million people to leave Vietnam as refugees. Also, thousands of Chinese were expelled. Many refugees left Vietnam in small boats, risking drowning and pirate attacks in the South China Sea. A large number of these refugees, who became known as *boat people,* eventually settled in the United States.

After the Vietnam War ended in 1975, Vietnam had troubles with neighboring countries. In 1978, Vietnamese troops helped the Cambodian Communists overthrow the Khmer Rouge and establish a new pro-Vietnamese government in Cambodia. Vietnamese troops then occupied Cambodia for 11 years.

In 1979, China, which had supported the Khmer Rouge, invaded and occupied Vietnam's northern border in retaliation. Continuing friction between Vietnam and China led to border clashes from time to time.

Vietnam depended on the Soviet Union for economic assistance but, by 1988, the Soviet Union had begun to reduce its aid to Vietnam. In an effort to attract investment and aid from Western nations, Vietnam introduced reforms that allowed some private enterprise.

Vietnamese leaders hoped that their withdrawal from Cambodia in September 1989 would open the way for foreign assistance. But the United States maintained the trade embargo imposed on Vietnam in 1975 following the Communist take-over.

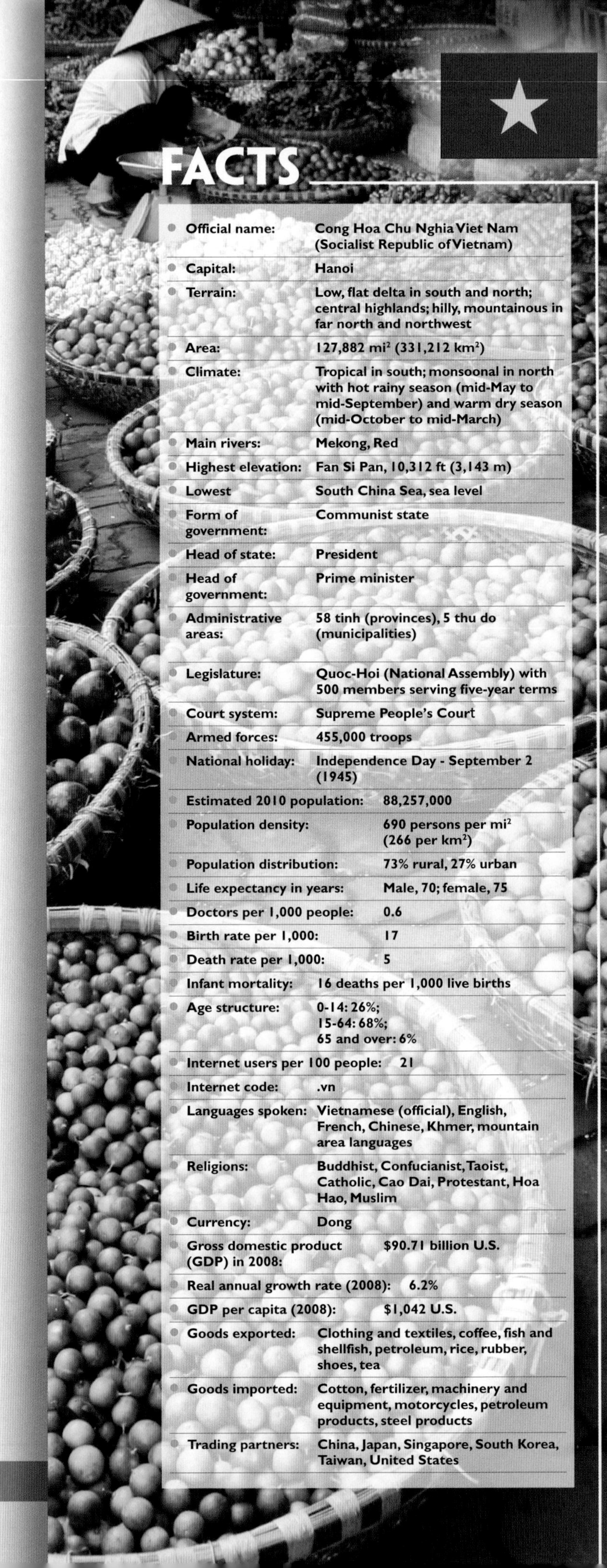

FACTS

Official name:	Cong Hoa Chu Nghia Viet Nam (Socialist Republic of Vietnam)
Capital:	Hanoi
Terrain:	Low, flat delta in south and north; central highlands; hilly, mountainous in far north and northwest
Area:	127,882 mi^2 (331,212 km^2)
Climate:	Tropical in south; monsoonal in north with hot rainy season (mid-May to mid-September) and warm dry season (mid-October to mid-March)
Main rivers:	Mekong, Red
Highest elevation:	Fan Si Pan, 10,312 ft (3,143 m)
Lowest	South China Sea, sea level
Form of government:	Communist state
Head of state:	President
Head of government:	Prime minister
Administrative areas:	58 tinh (provinces), 5 thu do (municipalities)
Legislature:	Quoc-Hoi (National Assembly) with 500 members serving five-year terms
Court system:	Supreme People's Court
Armed forces:	455,000 troops
National holiday:	Independence Day - September 2 (1945)
Estimated 2010 population:	88,257,000
Population density:	690 persons per mi^2 (266 per km^2)
Population distribution:	73% rural, 27% urban
Life expectancy in years:	Male, 70; female, 75
Doctors per 1,000 people:	0.6
Birth rate per 1,000:	17
Death rate per 1,000:	5
Infant mortality:	16 deaths per 1,000 live births
Age structure:	0-14: 26%; 15-64: 68%; 65 and over: 6%
Internet users per 100 people:	21
Internet code:	.vn
Languages spoken:	Vietnamese (official), English, French, Chinese, Khmer, mountain area languages
Religions:	Buddhist, Confucianist, Taoist, Catholic, Cao Dai, Protestant, Hoa Hao, Muslim
Currency:	Dong
Gross domestic product (GDP) in 2008:	$90.71 billion U.S.
Real annual growth rate (2008):	6.2%
GDP per capita (2008):	$1,042 U.S.
Goods exported:	Clothing and textiles, coffee, fish and shellfish, petroleum, rice, rubber, shoes, tea
Goods imported:	Cotton, fertilizer, machinery and equipment, motorcycles, petroleum products, steel products
Trading partners:	China, Japan, Singapore, South Korea, Taiwan, United States

In August 1990, the United Nations (UN) negotiated a peace plan calling for a supervised cease-fire in Cambodia and the establishment of a new coalition government. In 1991, U.S. officials outlined a four-step plan for normalizing relations with Vietnam. The U.S. embargo on trade was lifted in February 1994, and the two countries agreed to open liaison offices in Hanoi and Washington, D.C.

Also in 1994, many Vietnamese people who had emigrated after the Communist take-over returned. They brought expertise and money with them, and Vietnam's economy expanded significantly in the first half of the year. However, summer floods in the Mekong Delta Valley killed 300 people and reduced the nation's important rice crop.

Foreign investors, particularly from Taiwan and Hong Kong, began to invest in Vietnam. In 1995, Vietnam joined the Association of Southeast Asian Nations (ASEAN), a group of countries that promotes cooperation among its members. In 2000, Vietnam and the United States signed a trade agreement. This pact cleared the way for normal trade relations between the two countries for the first time since the Vietnam War. In 2006, the U.S. Congress passed a bill normalizing trade relations with Vietnam.

Vietnam is a country in Southeast Asia. It is bordered by China to the north and Laos and Cambodia to the west. Hanoi is the capital of Vietnam. Ho Chi Minh City, formerly named Saigon, is the largest city.

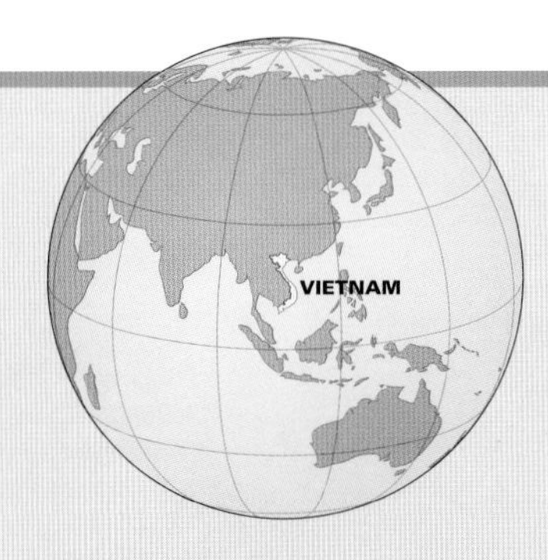

The tomb of Ho Chi Minh, president of North Vietnam from 1954 until his death in 1969, overlooks a street in Hanoi. Hanoi was once the capital of North Vietnam and is now the capital of unified Vietnam. Saigon, the former capital of South Vietnam, was renamed Ho Chi Minh City.

THE VIETNAM WARS

When World War II ended in 1945, Communist leader Ho Chi Minh and the Vietminh (Revolutionary League for the Independence of Vietnam) quickly took control of many areas of Vietnam, particularly the north. The Vietminh established the Democratic Republic of Vietnam (DRV), and the new government was also supported by many non-Communists who did not want to return to French colonial rule. Meanwhile, France regained control of Cochin China in southern Vietnam but was unable to put down all resistance. In December 1946, Vietminh forces attacked the French in Hanoi and the Indochina War began.

Although the French retained control of the cities in Vietnam, they were unable to gain power in the countryside. The Vietminh waged a guerrilla war in the hills and forests, shooting at French soldiers and escaping before the French could mobilize their forces.

An orphaned Vietnamese baby rides in a basket carried by an American soldier. Ground operations in South Vietnam had a devastating effect on civilian life.

American combat helicopters landed troops on "search and destroy" missions to seek out Viet Cong in the jungles and mountains of South Vietnam. Helicopters were also used to carry supplies and take the wounded to hospitals.

South Vietnam's countryside was scarred by U.S. bombing raids and by the spraying of plant-killing chemicals such as Agent Orange. The United States used these chemicals to reveal Communist hiding places in the jungle and to destroy their crops.

In 1953, France established a base at the village of Dien Bien Phu in northwestern Vietnam to disrupt Vietminh operations. However, in 1954, about 50,000 Vietminh soldiers destroyed the French base. France's defeat at the Battle of Dien Bien Phu ended its claim to the area.

The Geneva Accords, held between May and July 1954, arranged a cease-fire that ended the war between the French and the Vietminh. The agreements temporarily divided Vietnam into North Vietnam and South Vietnam and called for elections in 1956 to reunify the country. These elections were never held, however. President Ngo Dinh Diem of South Vietnam, fearing a Communist victory, would not let them take place. Both sides violated the cease-fire, and in 1957 South Vietnamese Communist guerrillas called the Viet Cong began the fighting that grew into the Vietnam War.

By 1960, the Viet Cong had about 10,000 troops and threatened to overthrow Diem's government. In support of the South Vietnamese government, the United States increased its military advisers in the country from about 900 in 1961 to more than 16,000 by 1963.

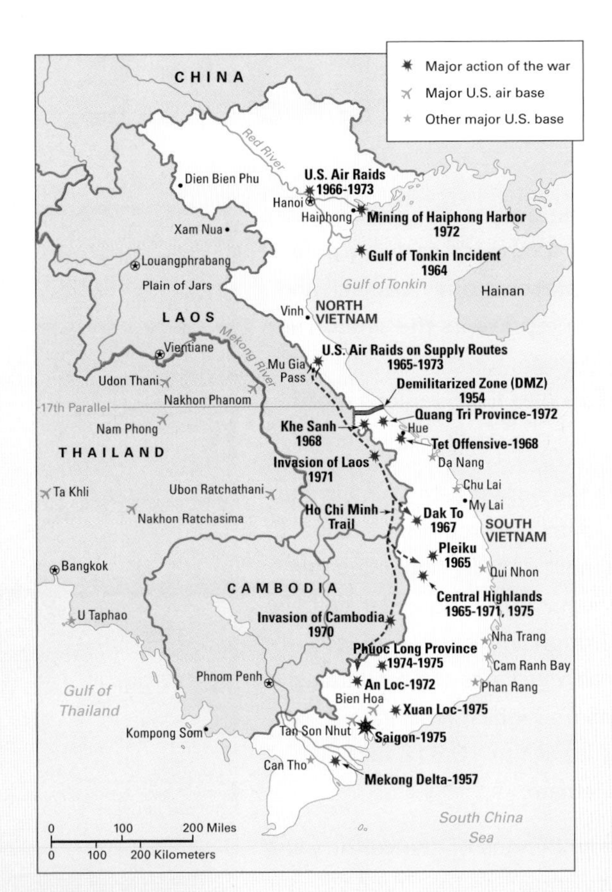

The Vietnam War was fought mainly in North and South Vietnam from 1957 to 1975. Troops also battled in Laos and Cambodia, and U.S. pilots flew missions from bases in Thailand.

North Vietnamese soldiers carry portraits of Ho Chi Minh through Saigon following the Communist victory in April 1975. Saigon, the capital of South Vietnam, was renamed Ho Chi Minh City after the take-over.

Then in 1964, President Lyndon B. Johnson announced that two U.S. destroyers had been attacked in the Gulf of Tonkin, off the coast of North Vietnam. As a result, Congress approved the Gulf of Tonkin Resolution, which allowed the president to take any action necessary against further aggression. Although the United States never declared war on North Vietnam, the 1964 resolution was used as the legal basis for increased U.S. military involvement. By 1969, the United States had more than 543,000 troops in Vietnam. The U.S. forces fought alongside about 800,000 South Vietnamese troops and some 69,000 allied soldiers.

The United States relied mainly on massive bombing raids in North Vietnam and ground missions in South Vietnam to combat the enemy. In contrast, the Communist forces used guerrilla tactics, including ambushes and hand-laid bombs. Their knowledge of the terrain, along with plentiful war supplies from the Soviet Union and China, aided this strategy.

As the fighting dragged on, many Americans at home—seeing the horrors of war revealed on television—began to oppose U.S. involvement in Vietnam. Antiwar demonstrations took place throughout the United States in the late 1960's and early 1970's. Opposition grew even more intense in 1970, when National Guard units killed four students and wounded nine others during a demonstration at Kent State University.

Heavy fighting and casualties on both sides led to peace negotiations in 1972, and in 1973, a cease-fire agreement was signed. North Vietnamese and Viet Cong troops resumed attacking South Vietnam, however, and the war continued—but without U.S. involvement. On April 30, 1975, when South Vietnam surrendered to North Vietnam, the war finally ended.

The Vietnam War took the lives of about 58,000 Americans and wounded about 365,000 more. Nearly 2 million North Vietnamese and South Vietnamese soldiers were killed in the fighting, as well as numerous Vietnamese civilians.

LAND AND PEOPLE

Vietnam's outline has reminded some people of two rice baskets hanging from opposite ends of a farmer's carrying pole. The Red River Delta in the north represents one "basket," and the Mekong River Delta in the south represents the other. The narrow stretch of land in central Vietnam forms the "pole." Geographers typically divide Vietnam into three regions: northern, central, and southern.

Northern Vietnam extends from the border with China in the north to about Thanh Hoa in the south. This region is dominated by the Red River Delta, the most densely populated center of agricultural production in Vietnam. The triangular delta is the heartland of Vietnamese civilization. The capital city of Hanoi is there. Northern Vietnam also includes the mountains of the north and northwest. Vietnam's highest mountain is Fan Si Pan, also spelled Phan Xi Pang. It rises to 10,312 feet (3,143 meters) in northwestern Vietnam.

Central Vietnam is the most mountainous of the country's regions. The Annamite Range, also known as the Truong Son mountains, dominates this area. The Central Highlands lie to the south. Poor soil makes farming difficult in central Vietnam. However, rich soil is available in the lowlands along the coast and a few plateaus in the Central Highlands.

Rafting goods down the Mekong River is an age-old method of transportation still carried on in Vietnam. The Mekong and its branches are important waterways throughout Southeast Asia.

The music of a bamboo flute lightens the tasks of farm children as they tend water buffaloes. These powerful animals pull a plow even when they are knee-deep in mud. They are highly valued by farmers.

The Mekong River Delta extends over much of southern Vietnam. The Mekong River flows from China through Southeast Asia into the South China Sea. The delta, Vietnam's chief agricultural area, is often referred to as the "rice bowl" of the country. Ho Chi Minh City, formerly named Saigon, is the region's major urban center and the country's economic hub.

Vietnam has dozens of ethnic groups. Over 85 percent of the people are Kinh—that is, ethnic Vietnamese—who are spread throughout the country. Minority ethnic groups live mainly in the mountain areas of

Workers harvest rice, Vietnam's chief crop. Most of the country's farmers practice wet-rice agriculture, in which rice is grown on irrigated paddies.

Street vendors sell vegetables in Ho Chi Minh City. The market places are crowded with peddlers who often display their goods on the sidewalks.

the country. The largest groups are the Tay, who live to the north and northeast of the Red River Delta, and the Tai, who live in scattered villages in the northwest and north-central interior. Other large minority groups include the Hmong, the Khmer, the Muong, and the Nung. A number of ethnic Chinese people, known as the Hoa, live mainly in the cities.

Most of the people in Vietnam live in villages and farm the land. Rural life has traditionally been built on strong family ties, where people set family interests above their own.

Traditionally, the eldest male served as head of the family, parents chose their children's marriage partners, and families honored their ancestors in special ceremonies. This way of life began to change in the late 1800's, when the French gained control of Vietnam and brought industry to the country. Many Vietnamese left the farms to work in the city factories.

However, the Vietnam War and its aftermath brought the most radical changes to Vietnamese life. To begin with, the war broke up families, as fathers, husbands, and sons left home to fight. In addition, the bombing of the countryside often drove rural people into the cities, where they learned Western customs from U.S. soldiers stationed there.

After the war, the Communist government controlled nearly every aspect of the people's lives. The government urged women to perform the same jobs as men, and it discouraged religion and the traditional honoring of ancestors.

The Vietnam War also disrupted the nation's educational system, and the government has had difficulty restoring it. Nevertheless, most of the Vietnamese people can read and write.

VIRGIN ISLANDS, BRITISH

Together with the U.S. Virgin Islands, the British Virgin Islands form the group known collectively as the Virgin Islands—part of the Leeward Islands group of the Lesser Antilles. The British Virgin Islands consist of nearly 60 islands, including Anegada, Jost van Dyke, Tortola, and Virgin Gorda islands, and their surrounding islets.

Many qualities of the British way of life are preserved on the British Virgin Islands. For instance, cars are driven on the left side of the road, as in the United Kingdom. Like most British people, most islanders are Protestants. They celebrate the birthdays of Queen Elizabeth II and the Prince of Wales as national holidays, and hotels serve afternoon tea—a British tradition. In the heat of the tropics, though, the tea is usually iced.

A tiny island near Beef Island can be reached only by boat. The British Virgin Islands' main airport is on Beef Island.

Sparkling blue waters and palm-shaded beaches—like this one at Deadman Bay on Peter Island—attract many tourists. The climate in the British Virgin Islands is warm and pleasant throughout the year.

The culture of the British Virgin Islands contrasts with the distinctive blend of African, Danish, and American influences found on the U.S. Virgin Islands. The British Virgin Islands are also far less developed than their U.S. counterparts. This may be partly due to the more favorable geographical position of the U.S. islands, which face the Caribbean Sea. The British Virgin Islands, on the other hand, look outward toward the vast Atlantic Ocean.

Island life

The British Virgin Islands have been under the British flag since 1672. Only 16 of the nearly 60 islands in the group are inhabited, and most of the people live on Tortola, the capital and largest island. About 23,000 people live on the islands. Many islanders work in the tourism industry, which generates nearly half of the territory's income. Other major economic activities include construction, the distillation of rum, financial services, and light industry.

Despite a steady stream of vacationers arriving from all corners of the globe, the tourism industry has so far left the islands relatively undisturbed. Resort developers have worked hard to prevent the destruction of natural habitats as they build hotel and restaurant facilities.

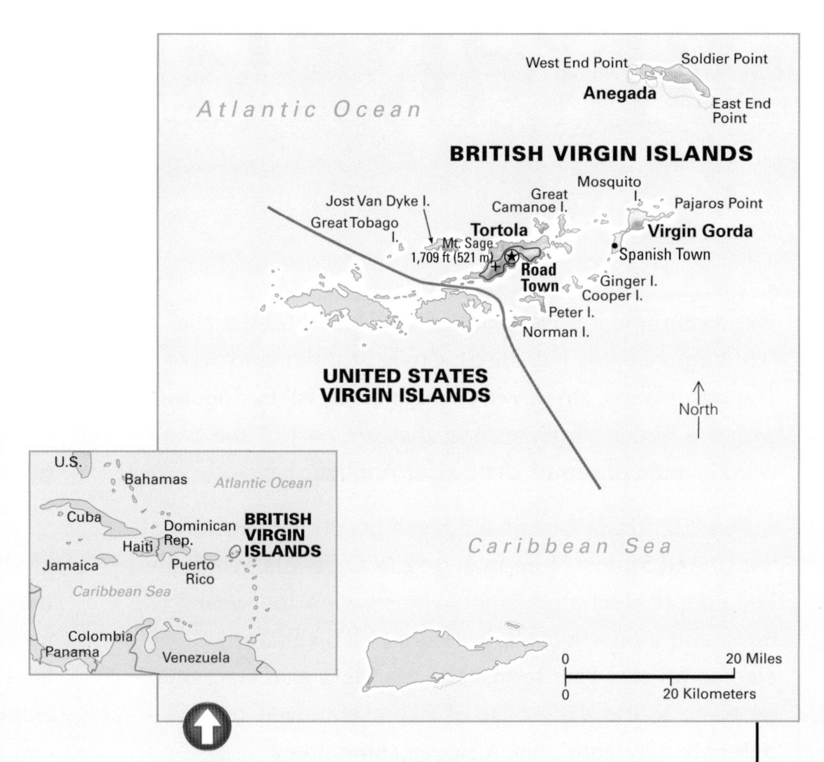

The British Virgin Islands lie near the western end of the Lesser Antilles. A channel called The Narrows separates the group from the U.S. Virgin Islands.

Most of the British Virgin Islands are hilly. The only exception of significant size is Anegada, a flat coral and limestone island surrounded by outstanding diving areas. In the underwater world of Horse Shoe Reef, adventurous divers can see hundreds of shipwrecks, as well as a variety of colorful marine life.

In the rural areas of the islands, most people live in small wooden huts with thatched roofs, and transportation is as basic as a ride into town on a donkey.

Tortola and Virgin Gorda

Tortola has a number of magnificent beaches, including those at Apple Bay and Belmint Bay. From the Road Town marina, studded with sleek and stylish yachts, many tourists take a boat to Dead Chest to see the famous coral reefs, or to Norman Island, said to be the inspiration for Robert Louis Stevenson's novel *Treasure Island*.

Nature lovers seeking peace and quiet often visit Virgin Gorda, the second largest of the British Virgin Islands. Beautiful beaches stretch along Virgin Gorda's southern coast, while steep cliffs line the north coast. Virgin Gorda also has magnificent natural caves. At Copper Mine Point, tourists may visit abandoned mines where Spaniards dug for copper, gold, and silver more than 400 years ago.

VIRGIN ISLANDS, U.S.

The easternmost possession of the United States, the U.S. Virgin Islands include St. Croix, St. John, and St. Thomas islands, along with many nearby islets. Together with the British Virgin Islands, they are part of the Leeward Islands group of the Lesser Antilles.

The U.S. Virgin Islands are a self-governing territory of the United States. More than 112,000 people live there. The people elect a governor, who serves a four-year term, and a one-house legislature of 15 senators, elected for two-year terms. The islanders also elect one delegate to the U.S. House of Representatives, but the delegate may vote only in House committees.

Christopher Columbus was the first European to discover the great natural beauty of the Virgin Islands, sighting the islands on his second voyage to the New World in 1493. Columbus was so captivated by the unspoiled appearance of the hills rising from the sea that he named the group the Virgin Islands, in memory of the 11,000 maidens who, according to legend, were guided by St. Ursula, a Roman Catholic saint.

St. Croix is the largest of the U.S. Virgin Islands, making up about two-thirds of the island group's area. St. Thomas is the second largest island, and St. John is the third largest. Fossils of ancient animals show that the islands were covered by the sea until volcanoes pushed the land up from the ocean floor. Except for St. Croix, the islands are rugged and hilly.

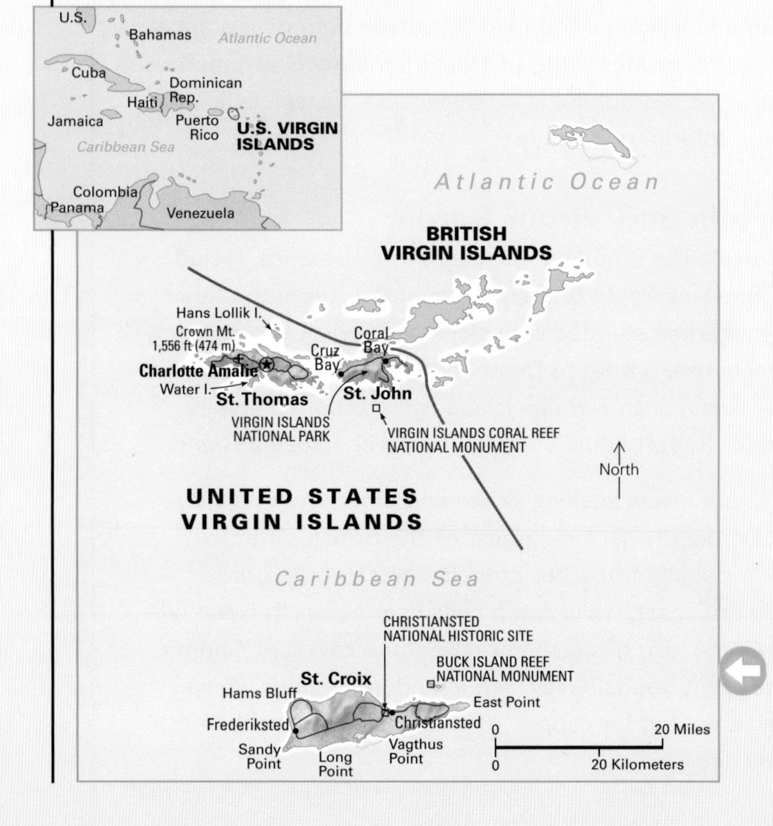

Early history

Although Columbus claimed the Virgin Islands for Spain, the Spaniards did not settle there. Instead, they used the islands as places to hide their treasure ships from pirates. It was not until 1625 that British and Dutch colonists established settlements on St. Croix. By then, the Carib Indians—the islands' original inhabitants—had either died or left the islands.

In the mid-1600's, Spaniards from Puerto Rico drove out the British and Dutch settlers, but the French defeated the Spaniards within 20 years. The French controlled St. Croix until 1733, when they sold it for $150,000 to the Danes, who had formally claimed St. Thomas in 1666. The Danish West Indies, which then included St. Croix, St. John, and St. Thomas, remained mainly under Danish control until 1917.

The total area of the U.S. Virgin Islands is 134 square miles (347 square kilometers). St. Croix, the largest island, covers 82 square miles (212 square kilometers); St. Thomas covers 27 square miles (70 square kilometers); and St. John is 19 square miles (49 square kilometers). Charlotte Amalie, on St. Thomas, is the capital of the U.S. Virgin Islands.

The composition of the rocks that form much of the land on the U.S. Virgin Islands suggests that volcanoes pushed the islands up from the ocean floor.

On Jan. 17, 1917, Denmark formally transferred ownership to the United States, which paid $25 million for the islands. Ten years later, the U.S. Congress passed a law making the people of the islands citizens of the United States.

Tourism

Tourism is the islands' major industry, with a large majority of the islanders employed in tourist-related jobs. Other islanders work in government jobs and in the rum distilleries, while some make their living as farmers.

On St. Croix, tourists enjoy exploring the towns of Christiansted and Frederiksted, where the charm of old Denmark can still be seen in pastel-colored buildings with red tile roofs. On St. Thomas, tourists enjoy duty-free shopping for imported luxury goods in the bustling port of Charlotte Amalie.

St. John is a natural paradise where hundreds of species of trees, flowers, and other plant life flourish in the Virgin Islands National Park, which covers about three-fourths of the island.

Cruise ships and yachts crowd the harbor of Charlotte Amalie on the island of St. Thomas. Beautiful scenery, fine beaches, and a warm, tropical climate make the U.S. Virgin Islands a favorite with vacationers, whose spending contributes significantly to the territory's economy.

WESTERN SAHARA

Western Sahara is a desert land on the northwestern coast of Africa, covering about 97,344 square miles (252,120 square kilometers). It lies between Morocco to the north, Algeria to the east, Mauritania to the east and south, and the Atlantic Ocean to the west.

Western Sahara is exactly what its name implies—the western end of the vast Sahara, a barren, rocky desert. Plant life is sparse except for patches of coarse grass and low bushes near the coast.

About 450,000 people live in Western Sahara—mainly Arabs and Berbers. The exact population is hard to determine because most of the people are nomads who move around the area in search of water and pasture for their herds of camels, goats, and sheep.

Western Sahara is rich in phosphates. These valuable chemicals are used as fertilizers and in making some detergents. Large deposits of phosphates have been found near the town of Bu Craa. Fish are plentiful off the Atlantic coast and provide a second valuable export.

The Saharawis, the inhabitants of Western Sahara are of mixed Arab and Berber descent. Since 1976, when the Polisario Front began fighting Morocco and Mauritania over control of the region, many Saharawis have settled in refugee camps in Algeria.

Children celebrate the founding of the Polisario Front. The Polisario Front is a nationalist group that opposes Morocco's claim to Western Sahara.

The Western Sahara is an area on the northwest coast of Africa. It lies between Morocco, Algeria, Mauritania, and the Atlantic Ocean.

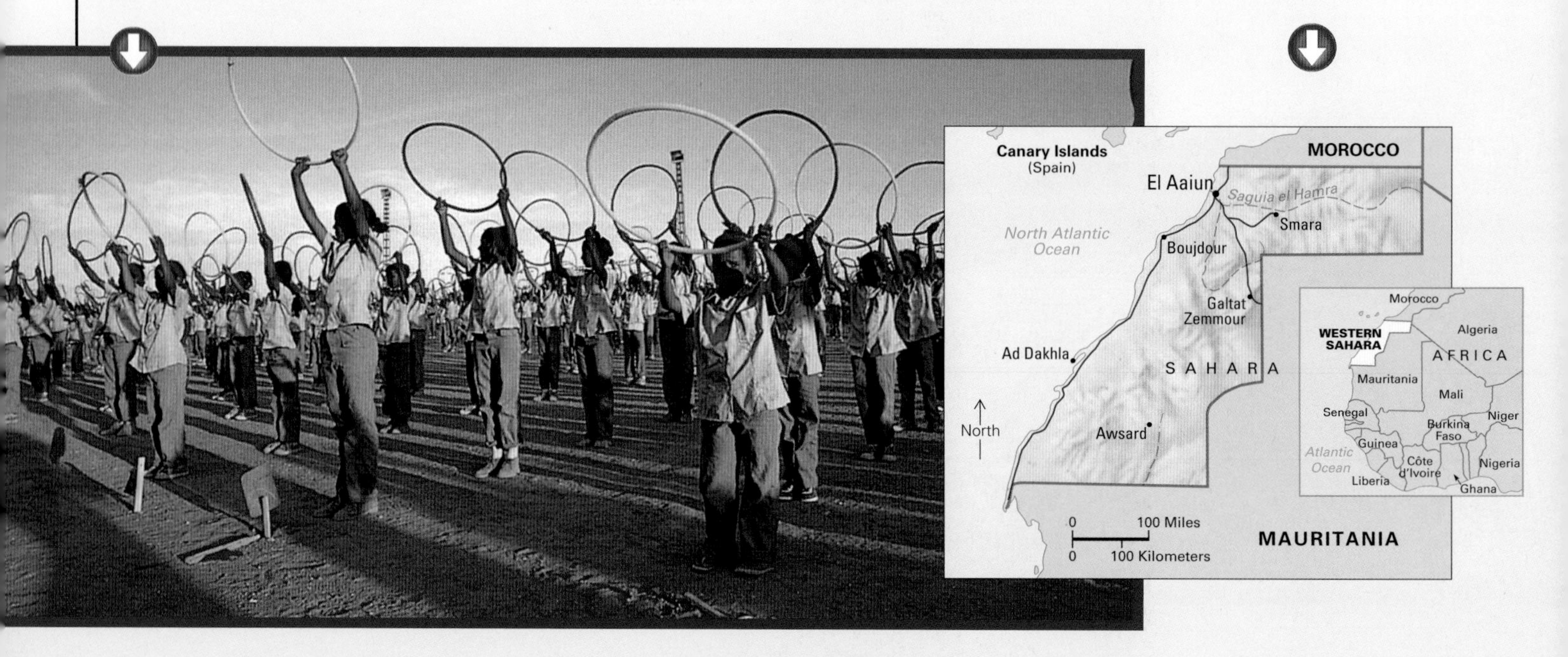

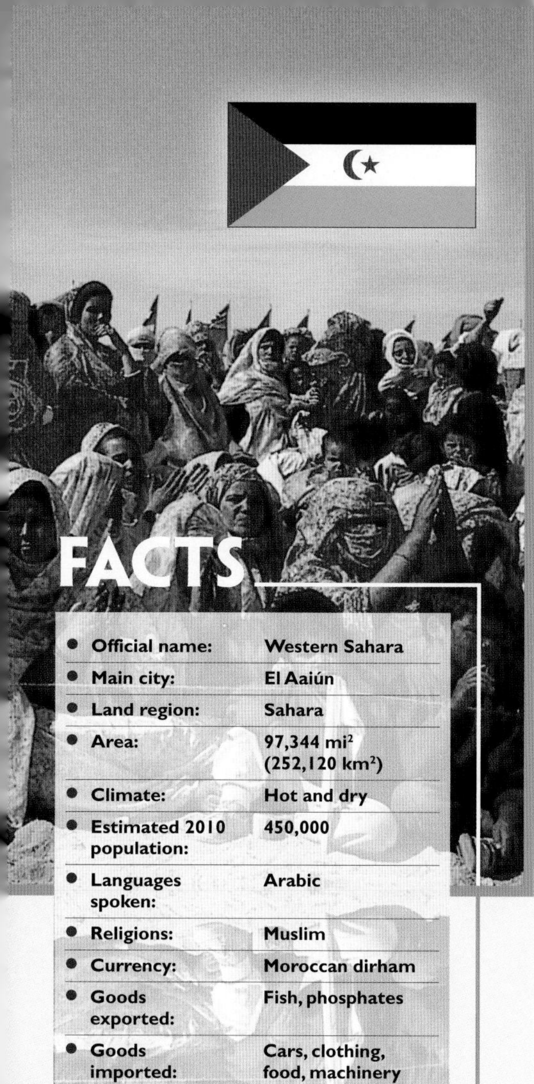

FACTS

• Official name:	Western Sahara
• Main city:	El Aaiún
• Land region:	Sahara
• Area:	97,344 mi^2 (252,120 km^2)
• Climate:	Hot and dry
• Estimated 2010 population:	450,000
• Languages spoken:	Arabic
• Religions:	Muslim
• Currency:	Moroccan dirham
• Goods exported:	Fish, phosphates
• Goods imported:	Cars, clothing, food, machinery

Berber women in the south of Western Sahara relax on beautiful carpets. A tent shelters them from the merciless heat of the Sahara.

The fight for Western Sahara

Today, possession of the territory of Western Sahara is disputed. Morocco claims the region, but its claim is opposed by some of the people who live in Western Sahara, as well as by Algeria.

The land was claimed by foreign powers as far back as 1509, when Spain claimed the area. Morocco then ruled the region from 1524 until Spain regained control in 1860. In 1958, Spain declared the area one of its provinces and called it the Province of Spanish Sahara.

In 1975, tens of thousands of Moroccans crossed their southern border into Spanish Sahara in support of King Hassan II of Morocco, who claimed the northern portion of Spanish Sahara for his country. Spain relinquished its claim to Spanish Sahara in 1976 and gave control of the region to both Morocco and Mauritania. The area then came to be called Western Sahara.

Morocco claimed northern Western Sahara, and Mauritania claimed the southern portion of the land. However, the country of Algeria protested these claims. Also, a group of Western Saharans called the Polisario Front demanded independence for the region.

Fighting broke out between the Polisario Front and Moroccan and Mauritanian troops. Algeria provided military aid for the Polisario Front. Later, Libya also aided the independence movement.

In 1979, Mauritania gave up its claim to the southern part of Western Sahara and withdrew its troops. Morocco then claimed the southern area as well.

Continuing conflict

During the 1980's, Moroccan troops built a 2,000-mile (3,200-kilometer) wall to seal off the most valuable areas of Western Sahara. A fortified defense line made of barbed wire and sandbanks and equipped with electronic sensors and minefields, the wall stretches from the Moroccan border southwest to the Atlantic coast.

A United Nations-supervised cease-fire between Polisario Front forces and Moroccan forces was declared in late 1991. The cease-fire plan also called for a referendum (direct vote) to determine whether Western Sahara would become independent or part of Morocco. However, disagreements between Morocco and the Polisario Front over voter eligibility have repeatedly delayed the vote.

YEMEN

Along the southern tip of the Arabian Peninsula lies the elbow-shaped country of Yemen. Its shores touch the Red Sea to the west and the Gulf of Aden to the south. The deserts of Saudi Arabia and Oman lie to the north and east. Although most of Yemen is hot and dry, the country's high northwestern interior is the most beautiful and cultivated part of the Arabian Peninsula.

Lowlands, highlands, and desert

In western Yemen, a coastal plain called the Tihamah lies along the shore of the Red Sea. In the south along the Gulf of Aden, the plain is mostly sand, with some fertile spots.

Both the Tihamah and the southern coastal plain are hot, with temperatures ranging from 68° to 130° F (20° to 54° C) in the Tihamah, and from 61° to 106° F (16° to 41° C) in Aden on the gulf. The Tihamah is humid, yet little rain falls on either lowland.

East of the Tihamah, cliffs rise up sharply from the plain. As much as 30 inches (76 centimeters) of rain falls on the cliffs each year, cutting into the rock and creating steep valleys. The cliffs form the western border of the highland region of Yemen.

The high interior is a land of broad plateaus lying 6,000 feet (1,800 meters) above sea level and surrounded by steep, rugged mountains rising more than 12,000 feet (3,000 meters) in the northwest. The altitude makes this region—called High Yemen—cooler and wetter than the coast.

As the highland region curves into southern Yemen, it becomes a dry, hilly plateau. Deep valleys called wadis cut into the plateau, providing some rich farmland.

East and north of the highlands, the land slopes to the stony desert called the *Rub al Khali* (Empty Quarter). The desert stretches past Yemen's undefined northern border into Saudi Arabia.

In Yemen's fertile highland region lies Sanaa, the capital. Sanaa is an ancient city—no one knows exactly when it was founded. A long wall encloses Sanaa and its distinctive buildings and mosques. Eight gates in the wall allow traffic to enter and leave the city. Many buildings in Sanaa are decorated with white plaster trim.

Many people in Yemen's cities live in mud-brick houses, and the town of Shibam has tall mud-brick buildings that seem to grow out of the land itself. Today, Western styles have begun to influence the country. Some of the city people live in modern apartments or in comfortable houses along wide streets.

An ancient land, young nation

The area we now call Yemen has a long history. Semitic people are thought to have invaded what is now northwestern Yemen about 2000 B.C. They introduced farming and building skills to the herders who were already living in the region. The area later grew rich as an ancient center of trade in frankincense and myrrh, as well as other luxury goods.

But Yemen itself is a fairly new country. From the 1700's until 1990, the territory that is now Yemen was made up of two countries—Yemen (Sanaa) and Yemen (Aden). The two Yemens had different histories: northwestern Yemen, or Yemen (Sanaa), was ruled by Muslim imams and was part of the Ottoman Empire. Southern and eastern Yemen, or Yemen (Aden), was a British protectorate. The two Yemens had somewhat different religious populations: Yemen (Sanaa) was largely divided between Shiite Muslims of the Zaydi sect and Sunni Muslims of the Shafii sect. Yemen (Aden) had mostly Shafii Sunnis. The two Yemens also had different economies and politics. Yemen (Sanaa) relied on agriculture and had an anti-Communist government, and Yemen (Aden) depended on its port at Aden and had a socialist government.

But the two Yemens also had much in common, and in 1990 they united to form the Republic of Yemen. Continuing tensions between North and South led to an eruption of civil war in May 1994. The North claimed victory on July 7, and President Ali Abdallah Salih declared the country reunited.

YEMEN TODAY

Yemen is a Middle Eastern country that wraps around the southern tip of the Arabian Peninsula. Its full name in Arabic, which is the country's official language, is Al-Jumhuriyah al Yamaniyah (the Republic of Yemen). Sanaa is Yemen's capital and largest city. Aden is the country's economic center.

The Republic of Yemen marks its date of birth as May 22, 1990. On that day, two separate nations—Yemen (Sanaa), which was also called North Yemen or Northern Yemen, and Yemen (Aden), which was also called South Yemen or Southern Yemen—announced their union.

A developing economy

Yemen remains one of the least developed countries in the world, and the nation depends on extensive foreign aid. Many of Yemen's young men leave the country to work and send some of their earnings home to their families.

Most Yemenis make a living by farming or herding, but Aden is an important port with a prosperous service economy. Ships of many nations use the port for refueling, repairs, and transferring cargoes. Aden's oil refinery processes oil shipped from other countries, mostly those on the Persian Gulf.

In the 1980's, petroleum deposits were found in several parts of the country, and development of Yemen's natural gas reserves began in 2005. However, only a small portion of Yemen's oil revenue has been devoted to building the economy and creating jobs.

Agriculture

Since the early 1980's, Yemenis have worked to turn desert areas into farmland with dams, irrigation, and other methods. Today, farming in the south and east is limited to the few valleys and

FACTS

Official name:	Al Jumhuriyah al Yamaniyah (Republic of Yemen)
Capital:	Sanaa
Terrain:	Narrow coastal plain backed by flat-topped hills and rugged mountains; dissected upland desert plains in center slope into the desert interior of the Arabian Peninsula
Area:	203,850 mi² (527,968 km²)
Climate:	Mostly desert; hot and humid along west coast; temperate in western mountains affected by seasonal monsoon; extraordinarily hot, dry, harsh desert in east
Main rivers:	N/A
Highest elevation:	Mount Hadur Shuayb, 12,336 ft (3,760 m)
Lowest elevation:	Arabian Sea, sea level
Form of government:	Republic
Head of state:	President
Head of government:	Prime minister
Administrative areas:	19 muhafazat (governorates); 1 administrative area
Legislature:	Shura Council with 111 members appointed by the president and House of Representatives with 301 members serving six-year terms
Court system:	Supreme Court
Armed forces:	66,700 troops
National holiday:	Unification Day - May 22 (1990)
Estimated 2010 population:	24,536,000
Population density:	120 persons per mi² (46 per km²)
Population distribution:	70% rural, 30% urban
Life expectancy in years:	Male, 61; female, 64
Doctors per 1,000 people:	0.3
Birth rate per 1,000:	41
Death rate per 1,000:	8
Infant mortality:	55 deaths per 1,000 live births
Age structure:	0-14: 45%; 15-64: 52%; 65 and over: 3%
Internet users per 100 people:	1.4
Internet code:	.ye
Language spoken:	Arabic
Religions:	Muslim including Sunni and Shia, Jewish, Christian, Hindu
Currency:	Yemeni rial
Gross domestic product (GDP) in 2008:	$27.10 billion U.S.
Real annual growth rate (2008):	3.2%
GDP per capita (2008):	$1,175 U.S.
Goods exported:	Mostly crude oil Also fish
Goods imported:	Chemicals, food, machinery, petroleum products, vehicles
Trading partners:	China, India, Switzerland, United Arab Emirates

oases with underground water for irrigation. In these areas, Yemeni farmers can grow three or four crops a year of barley, millet, sorghum, sesame, and wheat.

The plateaus and hills of the northwestern interior highlands are the most productive agricultural region of Yemen. Farmers raise food grains, such as barley, wheat, and a type of sorghum called *durra*. Yemenis grow a variety of fruits, including apricots, bananas, citrus fruits, dates, grapes, papayas, and pomegranates. Beans, lentils, onions, tomatoes, and other vegetables are grown in gardens at the edges of the highland villages and towns.

The leading cash crop, however, is *khat* (sometimes spelled *kat* or *qat*)—a woody highland shrub whose leaves contain a stimulant. When chewed, khat leaves produce a mildly intoxicating feeling of well-being. Groups of men and groups of women often meet separately in the afternoon to chew khat.

Coffee is another important cash crop for Yemenis. Coffee plants grow on terraces that are cut into the hills and watered by ancient aqueducts.

On the coast and on Socotra Island, some people live by fishing. Near the shore, Yemeni men spear fish from dugout canoes called sambugs; in deeper water, they net fish from single-sailed *dhows*.

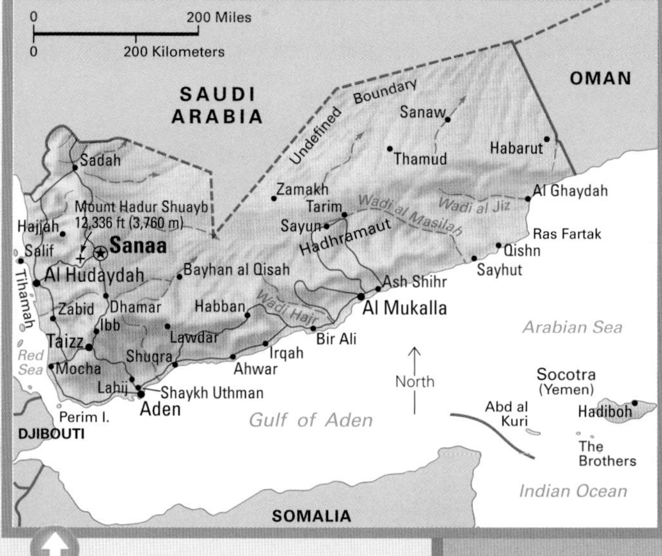

The Republic of Yemen came into being in 1990, with the union of Yemen (Sanaa) and Yemen (Aden). The country includes three offshore islands: Socotra in the Arabian Sea and Kamaran and Perim in the Red Sea.

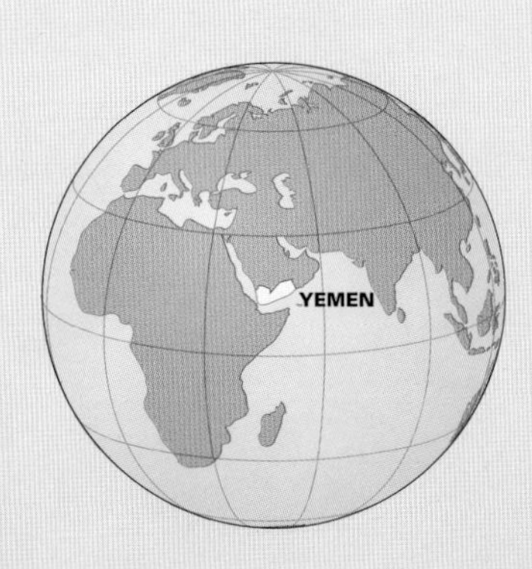

The houses of Sanaa, with their splendid balconies and brickwork, contribute to the city's distinctive architecture. Sanaa, now the capital of Yemen, was long the economic, political, religious, and educational center of Yemen (Sanaa).

HISTORY

In ancient times, Yemen was known as *Arabia Felix* (Fortunate Arabia)—a land of frankincense and myrrh, which yielded highly valued incense and perfumes. Yemen also lay along important trade routes between Europe, Asia, and Africa. Arab merchants there handled gems, spices, silks, and other exotic goods. Cities, castles, temples, and dams were built during this period, and farms were irrigated. In the 900's B.C., the Queen of Sheba ruled one of several kingdoms that flourished in the area.

Ruins of the Marib Dam are evidence of the ancient civilization that flourished in Yemen. Built about 500 B.C., the dam provided enough water to irrigate an area that could have supported as many as 300,000 people.

But the region's prosperity ended in the A.D. 100's. Local tribal chiefs warred among themselves, and Abyssinia (now Ethiopia) conquered the land for a short time in the A.D. 300's. For hundreds of years afterward, Yemeni tribes and religious groups fought with one another, as well as with would-be invaders.

The most important cultural and political event in the history of the region occurred in the 600's, when Islam was introduced to the people. By the end of the 800's, most Yemenis were Muslims.

The adoption of Islam did not solve all the tribal conflicts, however, partly because tribes in the region belonged to different Muslim sects. To a large degree, the tribes remained split socially, politically, and religiously. The history of the region split too.

Yemeni women wash clothes in a rocky stream using a method that may have remained unchanged for centuries. Yemen no longer enjoys the position of wealth and importance it held in ancient times.

Beginning in 897, a Muslim *imam* began to serve as the religious and political leader of northwestern Yemen. In 1517, this area fell under the control of the Ottoman Empire, which was centered in what is now Turkey. However, the local imams kept considerable power. In 1924, the Treaty of Lausanne freed northwest Yemen from the Turks.

In 1962, a group of military officers supported by Egypt overthrew the imam and set up a republic. The imam's forces—called *royalists*—were supported by Saudi Arabia and fought from bases in the mountains to regain control. The fighting ended in 1970, when the republicans set up a government that included both republicans and royalists.

Meanwhile, southern and eastern Yemen had fallen under Western influence. In 1839, the United Kingdom had seized the city of Aden in southern Yemen after people of the town looted a wrecked British ship. Aden then became an important refueling stop for British ships on their way to India via the Suez Canal.

The British wanted to protect Aden from the rest of southern Yemen, which claimed the town. Therefore, the United Kingdom extended its control to the tribal states in the surrounding region and signed treaties with the tribal chiefs, promising protection and aid in return for loyalty. The area came to be known as the Aden Protectorate.

In 1959, six tribal states in the protectorate formed a federation, and the United Kingdom promised the federation independence in the future. By 1965, Aden and all but four of the tribal states had joined the federation.

However, the British had trouble setting up a representative government to rule the federation after independence. Both the radical Arab nationalists in Aden and the tribal chiefs in the protectorate wanted control. The radicals launched terrorist attacks against both the British and the chiefs. Two radical groups—the National Liberation Front (NLF) and the Front for the Liberation of Occupied South Yemen (FLOSY)—also fought each other for control.

In 1967, British troops finally withdrew, and the NLF, the most powerful group, formed a government for the new country, called Yemen (Aden). Later, the NLF and other groups reorganized as the Yemeni Socialist Party (YSP).

National leaders then formed ties with Communist countries, which led to fighting with northwestern Yemen, then called Yemen (Sanaa). Anti-Communist army officers took over Yemen (Sanaa) in 1974. Relations between the Yemens began to improve in the 1980's, and in 1990, the two were at last united. Disputes over power sharing and oil resources led to civil war in May 1994, but the fighting ended in July with victory for the North. The country remained one.

The rugged hills of northern Yemen once supported forests that yielded the rare and fragrant resins called frankincense and myrrh. Ancient peoples prized these resins for use in medicines, cosmetics, and incense.

PEOPLE AND GOVERNMENT

Most of the people of Yemen are Arab Muslims, but they belong to different tribes and different religious sects. Small groups of Indians, Pakistanis, and Africans also live in Yemen.

Way of life

Most of Yemen's people are farmers or herders. Some make their living by fishing in coastal waters. In the desert, a few people still live as nomadic herders. They travel constantly in search of water and food for their sheep and goats. Because of Yemen's lack of economic development, many Yemenis depend on wages sent home by family members working in neighboring oil-rich lands. Some areas of Yemen have no schools, and less than half of Yemenis 15 years of age or older can read and write.

The basic foods of Yemen's people are rice, bread, vegetables, lamb, and fish. A spicy stew called *salta* is the national dish.

As in other developing countries, city life in Yemen differs from rural life. Some city people reside in modern houses or apartment buildings. Many others live in one-story brick houses. Many wear Western-style clothing

Some farm families live in towns, such as Sayun, that have mud-brick houses three or four stories high. Others live in small villages close to the land they farm. Near the Red Sea coast, many people live in straw huts. Traditional Arab clothing is common in rural areas. The men's garments include cotton breeches or a striped *futa* (kilt). Many men wear skullcaps, turbans, or tall, round hats called *tarbooshes.*

Many Yemenis are craft workers in small one-room shops. Yemeni craft workers have been noted for their textiles, leatherwork, and ironwork since ancient times. Many such goods are still made by hand. Yemenis dye and weave beautiful cloth and make rope, glassware, wooden chests, *jambiyas* (daggers), jewelry, brassware, harnesses, saddles, and pottery. Some sell their goods in village bazaars, while others go to Aden's market district, where each trade has its own street.

A forbidding landscape of rock and sand makes up the northern interior of Yemen. A few nomadic herders wander through the desert. Most of the men own nothing but their clothes and their curved daggers called jambiyas. The women are unveiled, and many are tattooed with tribal marks.

A teacher instructs a class of eager children in a Yemeni school. Education was stressed more in Yemen (Aden), which had been controlled by the British, than it was in Yemen (Sanaa).

A jambiya salesman displays his wares. In the past, the kind of jambiya a man owned and the position in which it was worn indicated the tribe to which he belonged.

A dark veil, embroidered with silver, hides the face of a Yemeni woman. Especially in cities and towns, women are veiled and wear black head-to-foot gowns.

Conflict and division

Tribal divisions have played an important role in Yemen since ancient times. The Arab tribes' rivalries and their desire for self-rule often led to conflicts.

Most of Yemen's people are Muslims of the Zaydi or Shafii sects. The Zaydis live in northwestern Yemen. They have long been associated with the government and the military. The Shafii sect has a powerful merchant class. The division between the politically powerful Zaydis and the wealthy Shafiis has caused bitterness between them.

East meets West as Yemeni men sport Western-style jackets over long tribal robes. Many Yemenis, however, prefer traditional Arab clothing.

A united government

Before the two Yemens merged, each had its own national legislature. In 1990, the two legislative assemblies elected a five-member ruling council for the united country. The legislatures then dissolved themselves and reformed as the House of Representatives. The House has 301 members. Yemen's parliament also includes a Shura Council, with 111 members appointed by the president.

Tensions over power-sharing and oil resources led to civil war in May 1994. The South declared its independence and attacked the Northern capital of Sanaa. The war ended when the South's main city, Aden, fell on July 7. As many as 12,000 people died.

Civil and political violence has continued to plague Yemen. In 1998, 12 tourists were kidnapped, and 4 were murdered. Terrorists bombed the U.S. warship *Cole* at Aden in 2000, killing 17 U.S. sailors. In 2008, a suicide bombing at the U.S. embassy in Sanaa killed 19 people. In 2004, a Zaydi rebellion erupted in the far north, and violence escalated into open warfare in 2009. The two sides signed a peace treaty in February 2010, but sporadic violence has continued.

In January 2011, antigovernment protests broke out. In November, President Ali Abdullah Salih, who had ruled Yemen for 33 years, transferred power to his vice president. Vice President Abdel Rabbo Mansour Hadi was elected president in February 2012, in an election in which he was the only candidate.

ZAMBIA

The Republic of Zambia in south-central Africa took its name from the great Zambezi River, which forms most of its southern border. Once a British protectorate called Northern Rhodesia, Zambia became an independent nation in 1964.

Government

The president is the most powerful official in Zambia. The people elect a president as well as most of the members of the National Assembly, the country's legislature. A few members are appointed by the president, who also appoints a vice president and a cabinet to help run the government.

Kenneth Kaunda, who led the movement that resulted in independence, served as president from 1964 until 1991. In 1972, the United National Independence Party became the only legal political party in Zambia. However, in 1990 a constitutional amendment permitted the formation of opposition parties.

In a multiparty election in 1991, voters elected Frederick Chiluba of the Movement for Multiparty Democracy (MMD) as Zambia's new president. He was reelected in 1996. Levy Mwanawasa, also of the MMD, was elected president in 2001 and reelected in 2006. When Mwanawasa died in office in 2008, his vice president, Rupiah Banda, was elected to serve out his term. In 2011, voters chose Michael Sata of the Patriotic Front as president.

Economic challenges

Zambia faced economic difficulties in the 1970's and 1980's, due partly to strained relations with neighboring Rhodesia (now Zimbabwe). Rhodesia, also a former British protectorate, had declared its independence without British consent. Rhodesia's government was ruled by whites, even though blacks greatly outnumbered them. Relations between Zambia and Rhodesia deteriorated over the Rhodesian government's refusal to give the African majority a greater voice.

In 1973, Rhodesia prohibited Zambia from shipping goods across its territory, thus eliminating one of Zambia's main outlets to the sea. Rhodesia soon lifted

FACTS

Official name:	Republic of Zambia
Capital:	Lusaka
Terrain:	Mostly high plateau with some hills and mountains
Area:	290,585 mi² (752,612 km²)
Climate:	Tropical; modified by altitude; rainy season (October to April)
Main rivers:	Zambezi, Luangwa, Kafue
Highest elevation:	Unnamed location in Mafinga Hills, 7,549 ft (2,301 m)
Lowest elevation:	Zambezi River, 1,079 ft (329 m)
Form of government:	Republic
Head of state:	President
Head of government:	President
Administrative areas:	9 provinces
Legislature:	National Assembly with about 160 members serving five-year terms
Court system:	Supreme Court, High Court
Armed forces:	15,100 troops
National holiday:	Independence Day - October 24 (1964)
Estimated 2010 population:	12,689,000
Population density:	44 persons per mi² (17 per km²)
Population distribution:	64% rural, 36% urban
Life expectancy in years:	Male, 38; female, 38
Doctors per 1,000 people:	0.1
Birth rate per 1,000:	40
Death rate per 1,000:	21
Infant mortality:	101 deaths per 1,000 live births
Age structure:	0-14: 46%; 15-64: 52%; 65 and over: 2%
Internet users per 100 people:	5
Internet code:	.zm
Languages spoken:	English (official), Bemba, Kaonda, Lozi, Lunda, Luvale, Nyanja, Tonga, other indigenous languages
Religions:	Christian 50%-75%, Muslim and Hindu 24%-49%, indigenous beliefs 1%
Currency:	Zambian kwacha
Gross domestic product (GDP) in 2008:	$14.32 billion U.S.
Real annual growth rate (2008):	5.8%
GDP per capita (2008):	$1,169 U.S.
Goods exported:	Mostly copper Also cotton, food, tobacco
Goods imported:	Chemicals, machinery, petroleum products, vehicles
Trading partners:	China, Democratic Republic of the Congo, South Africa, Switzerland

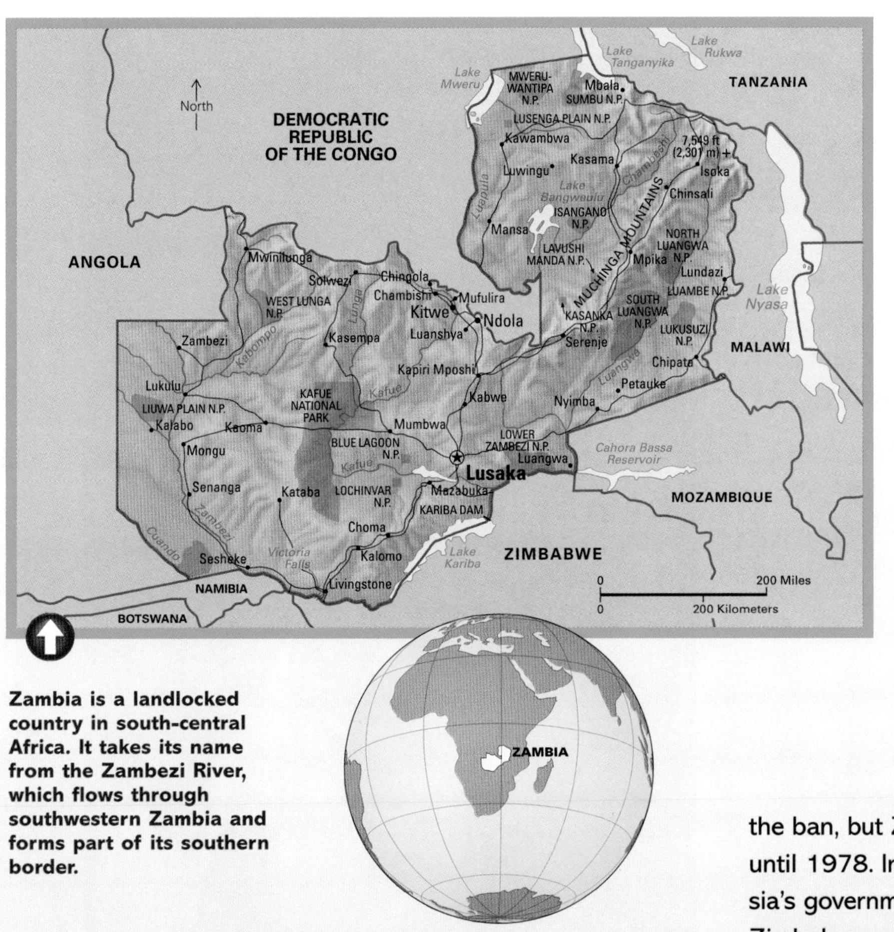

Zambia is a landlocked country in south-central Africa. It takes its name from the Zambezi River, which flows through southwestern Zambia and forms part of its southern border.

The mining and refining of copper is the basis of Zambia's economy and accounts for much of the country's export earnings.

the ban, but Zambia refused to ship goods across Rhodesia until 1978. In 1980, when blacks gained control of Rhodesia's government and changed the name of the country to Zimbabwe, relations between the two countries improved.

Zambia was also hurt economically when the price of copper dropped on the world market. Zambia ranks as one of the world's largest producers of copper, and the mineral accounts for about half of the country's export earnings. In addition, the production of copper products is the country's most important manufacturing activity. Without copper, Zambia would be one of the poorest countries in Africa.

The economy depends heavily on the mining of other minerals as well. Under President Kaunda, the government controlled the economy. It tried to lessen Zambia's dependence on mining and to increase the importance of other economic activities. However, these measures largely failed, and Zambia's economy weakened further. The government has since sought to privatize industries and introduce market economics in an effort to strengthen the economy.

Victoria Falls, called Mosi oa Tunya (smoke that thunders) by local people, is one of Africa's most awesome sights. The falls lies between Zambia and Zimbabwe on the Zambezi River.

LAND AND PEOPLE

The land of Zambia is mostly a plateau lying about 4,000 feet (1,200 meters) above sea level. Because of its altitude, the country has a milder climate than might be expected in tropical Africa. Trees and bushes cover most of the country's relatively flat expanse.

The plateau is broken by the Muchinga Mountains, which rise 7,000 feet (2,100 meters) in the northeast. In the southwest, a broad, sandy plain lies on either side of the Zambezi River. Every year during the rainy season, the river waters flood the plain.

The wet season lasts from November through April, and violent storms swell the rivers of the country by March. Northern Zambia gets about 50 inches (130 centimeters) of rainfall a year, while the south gets 20 to 30 inches (51 to 76 centimeters). Because of this decrease in rain from north to south, trees in the south are smaller, and large open areas are found there.

Many of Zambia's people farm the land for a living, but Zambian farmers cannot grow enough food to feed the country's rapidly growing population. Only a small percentage of the country's land area is cultivated, but a much larger area is suitable for farming. Every year, Zambians clear many square miles of forests to create much-needed farmland.

A spirit dancer, wearing a bright red mask framed by a black wig, is a member of one of Zambia's animistic groups. Animists believe all things in nature have souls.

The Zambian landscape is made up mostly of wide expanses of trees and bushes. This stretch of land lies along the course of the Luangwa River, where water levels fall considerably during the dry season.

Roman Catholic Zambians attend a mass. The majority of Zambians are Christians, but traditional African beliefs and religious rituals are still widely practiced.

In the *bush*, or rural areas, where most Zambians live, the people plant their crops in November and December, at the beginning of the rainy season. In these remote parts of the country, life goes on much as it has for hundreds of years. The people live in villages of circular, grass-roofed huts and go out to the surrounding land to raise their food crops. Corn is the country's most important farm product and the people's main food. A favorite dish is *nshima*, a thick corn porridge. Other crops include cassava, coffee, millet, sorghum, sugar cane, and tobacco.

The development of mining in Zambia has drawn thousands of jobseekers to mining towns. Other Zambians live and work in the capital city of Lusaka, where the government employs many people. Lusakan factory workers help produce such goods as beverages, cement, food products, furniture, shoes, textiles, and tobacco. Nevertheless, less than 40 percent of all Zambians live in urban areas.

Most Zambians are black Africans who speak Bantu languages. There are more than 70 ethnic groups represented and eight major local languages spoken in Zambia. Many Zambians also speak English, the country's official language.

The majority of Zambians are Christians, but traditional African beliefs strongly influence the village people. Traditional herbal medicine is still practiced in some rural areas, and old customs such as polygyny (one man marrying several wives) and bride price (a man paying parents in order to marry their daughter) are still followed. However, these traditions are slowly dying out in the towns.

The vast majority of Zambian children attend elementary school, but only a fifth of them go on to secondary school. Those who graduate may attend the University of Zambia or one of the nation's trade or technical schools.

Shouldering their tools, Zambian boys return to their village after a hard day's work on their families' farms. Zambia's soil is generally poor, and the people make only a bare living off the land.

Cairo Road is the main thoroughfare of Lusaka, the capital of Zambia. The road is the principal business, retail, and services center of the city.

ZIMBABWE

The country of Zimbabwe, once known as Rhodesia, lies on a high rolling plateau in tropical southern Africa. Most of Zimbabwe is 3,000 to 5,000 feet (910 to 1,500 meters) above sea level. This high altitude helps create a pleasant climate.

Landscape

Zimbabwe's beautiful scenery includes the famous Victoria Falls on the Zambezi River, which runs along the country's northern border. About halfway between its mouth and its source, the Zambezi—about 1 mile (1.6 kilometers) wide at this point—drops suddenly into a narrow, deep chasm. The mist and spray created by this magnificent waterfall can be seen from several miles away. Because of this permanent cloud and the constant roar of the falling water, the people of the area named the falls *Mosi oa Tunya* (smoke that thunders). When Scottish explorer David Livingstone first sighted the falls in 1855, he named it after Queen Victoria of the United Kingdom.

A canyon about 40 miles (64 kilometers) long permits the water to flow out of the chasm. The height of Victoria Falls varies from 256 feet (78 meters) at the right bank to 355 feet (108 meters) in the middle.

Most rivers in Zimbabwe flow away from the center of the country—either northwest into the Zambezi or southeast—because the High Veld, a central grassy plateau, crosses Zimbabwe from northeast to southwest. The Middle Veld lies on either side of the High Veld. The Low Veld consists of sandy plains in the basins of the Zambezi, Limpopo, and Sabi rivers.

Women and children hoe corn on a Zimbabwe farm. Agriculture is the country's largest employer. In addition to corn, Zimbabwe produces livestock products and wheat for domestic use. The chief agricultural exports include cotton, sugar, and tobacco.

The Chimanimani Mountains rise along Zimbabwe's eastern border. Chimanimani National Park, one of the country's land and animal reserves, lies in this area.

Miners play an important role in Zimbabwe's economy. Minerals, especially gold, make up much of the country's exports.

A masked dancer invokes spirits at a religious gathering near Harare. Many people in Zimbabwe follow Christianity blended with traditional African beliefs, such as animism—the belief that everything in nature has a soul—and ancestor worship.

The people and their work

Most of the High Veld grasslands are owned by white farmers, even though whites make up only about 1 percent of the population. Many blacks work on these commercial farms. Crops include coffee, corn, cotton, sugar, sunflower seeds, tea, tobacco, and wheat. Cattle are also raised on large ranches.

Most black Africans, who make up about 98 percent of Zimbabwe's population, are farmers who raise only enough food for their families. Their main crop is corn, which they pound into flour to make a dish called *sadza*.

About 1 percent of Zimbabwe's population is made up of Asians and *Coloureds* (people of mixed ancestry). Most of the whites, Asians, and Coloureds live in the nation's urban areas. Harare, Zimbabwe's capital and largest city, is a trading center for products raised on the fertile High Veld. Modern high-rise hotels and office buildings dominate downtown Harare. Crowded slums as well as upper-class suburbs lie on the outskirts of the city.

Most blacks in Zimbabwe live in rural areas, in thatched huts. The Shona people—often called the Mashona—make up the largest African ethnic group and speak a language called chiShona. The Ndebele—often called the Matabele—are the second largest ethnic group, and their language is isiNdebele.

English remains the official language of Zimbabwe and the language of business. However, since the black majority gained control of Zimbabwe's government in 1980, many whites have left the country.

Abundant mineral wealth

Europeans originally came to Zimbabwe because of its mineral wealth. Cecil Rhodes, a British-born businessman, gained mineral rights from the Ndebele in 1888. Reports of gold brought more Europeans to the area in the 1890's.

Today, Zimbabwe is still an important producer of gold, as well as asbestos, coal, nickel, and other minerals. A smelter at Kwekwe (formerly Que Que) removes iron from ore mined in the area, and coal comes from the Hwange (formerly Wankie) region. Zimbabwe also has deposits of chromite, copper, tin, and gems. Although Zimbabwe must import oil, the huge Kariba Gorge hydroelectric complex on the Zambezi River supplies electricity to most of the country.

ZIMBABWE TODAY

Zimbabwe has had a troubled, and sometimes violent, political history. In the late 1800's, the land was part of Rhodesia, a territory named after British businessman Cecil Rhodes. Although the vast majority of Rhodesia's people were black Africans, whites controlled the government. Southern Rhodesia (now Zimbabwe) became a colony separate from Northern Rhodesia in 1923.

In the 1960's, black Africans in Southern Rhodesia began asking for a greater voice in the government. The white government eventually banned two black African parties.

In 1964, when Northern Rhodesia became the independent nation of Zambia, Southern Rhodesia became known simply as Rhodesia. The white government demanded independence for Rhodesia, but the United Kingdom refused. When Prime Minister Ian Smith of Rhodesia declared independence in 1965, the United Kingdom claimed the act was illegal and banned trade with Rhodesia.

In 1969, Rhodesian voters—mostly whites —approved a new constitution designed to prevent the black majority from gaining control. Rhodesia was declared an independent republic in 1970, but other countries did not recognize the declaration. Most continued to apply economic and political pressure to end white rule. Black nationalists began fighting the government within Rhodesia.

These actions finally forced Smith to agree to hand some political power to blacks. The first black-majority government was elected in April 1979, with Abel T. Muzorewa as prime minister. However, many blacks rejected this government because they felt it did not truly represent the majority. Guerrilla violence continued until an agreement was reached to create a new government.

In February 1980, the Zimbabwe African National Union-Patriotic Front (ZANU-PF) party won a majority of seats in the country's Parliament. Robert Mugabe, the party's leader, became prime minister. On April 18, 1980, the United Kingdom recognized the independence of the new nation, now called Zimbabwe. Most other nations recognized the new country and lifted trade sanctions.

FACTS

Official name:	Republic of Zimbabwe
Capital:	Harare
Terrain:	Mostly high plateau with higher central plateau; mountains in east
Area:	150,872 mi² (390,757 km²)
Climate:	Tropical, moderated by altitude; rainy season (November to March)
Main rivers:	Zambezi, Limpopo, Sabi, Shangani
Highest elevation:	Mount Inyangani, 8,514 ft (2,595 m)
Lowest elevation:	Junction of the Runde and Save rivers 531 ft (162 m)
Form of government:	Parliamentary democracy
Head of state:	President
Head of government:	Prime minister
Administrative areas:	8 provinces, 2 cities with provincial status
Legislature:	Parliament consisting of the House of Assembly with 210 members serving five-year terms and the Senate with 93 members serving five-year terms
Court system:	Supreme Court, High Court
Armed forces:	29,000 troops
National holiday:	Independence Day - April 18 (1980)
Estimated 2010 population:	13,733,000
Population density:	91 persons per mi² (35 per km²)
Population distribution:	63% rural, 37% urban
Life expectancy in years:	Male, 43; female, 43
Doctors per 1,000 people:	0.2
Birth rate per 1,000:	31
Death rate per 1,000:	18
Infant mortality:	59 deaths per 1,000 live births
Age structure:	0-14: 40%; 15-64: 56%; 65 and over: 4%
Internet users per 100 people:	11
Internet code:	.zw
Languages spoken:	English (official), chiShona, isiNdebele, numerous minor tribal dialects
Religions:	Syncretic (part Christian, part indigenous beliefs) 50%, Christian 25%, indigenous beliefs 24%, other 1%
Currency:	United States dollar
Gross domestic product (GDP) in 2008:	N/A
Real annual growth rate (2008):	N/A
GDP per capita (2008):	N/A
Goods exported:	Cotton, food, iron and steel, nickel and other metals, tobacco
Goods imported:	Chemicals, corn, machinery, petroleum and petroleum products, vehicles
Trading partners:	Botswana, China, Mozambique, South Africa

Children search for food in garbage heaps in a poor township. During the 2000's, a combination of drought and government incompetence and corruption led to widespread food shortages, and millions of people faced starvation.

Zimbabwe is a landlocked country on a high plateau in southern Africa. Zimbabwe lies in the tropics but has a pleasant climate because of the high altitude. Zimbabwe is a leading mineral producer.

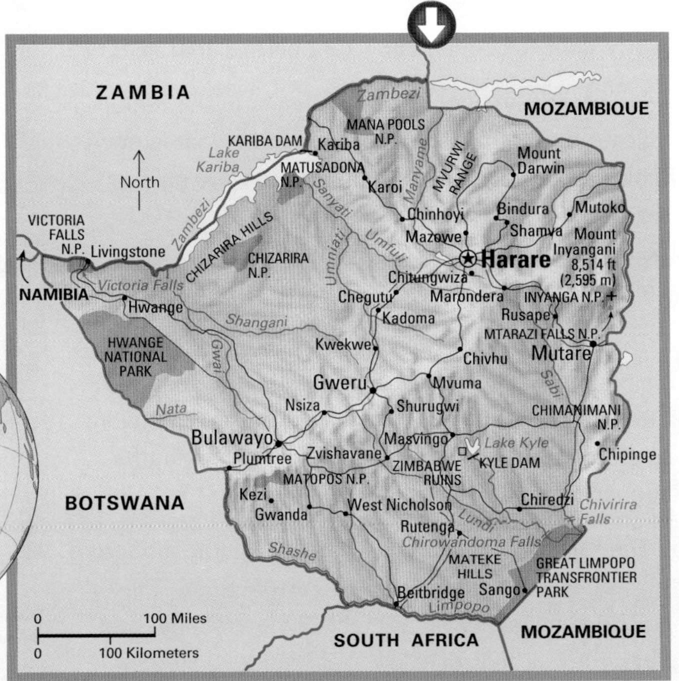

In 1981, fighting broke out between two black ethnic groups—the Ndebele and the Shona. Most of the fighting ended in 1983.

Mugabe was elected to the new office of executive president in 1987 and was reelected several times. But in 2002, the opposition party, the Movement for Democratic Change (MDC), rejected Mugabe's reelection, saying their party had been the target of political violence and intimidation.

In the early 2000's, drought conditions led to widespread shortages in Zimbabwe, and millions of people faced starvation. In 2002, Mugabe's government ordered about 3,000 white farmers to leave their land. The government then began seizing the white-owned farms for redistribution to black people. Many white farmers disobeyed the order, and hundreds of them were arrested.

In March 2008, Zimbabwe held presidential and parliamentary elections. For the first time since 1980, ZANU-PF did not win a majority of seats. Morgan Tsvangirai of the MDC had the most votes, but not enough to win the election. A runoff presidential election between Tsvangirai and Mugabe was held, but Tsvangirai withdrew, claiming that the election would not be fair. Mugabe was elected president. In September, a power-sharing agreement was signed by both parties that created the post of prime minister, to be held by Tsvangirai. Tsvangirai took ofice in 2009. However, tensions between the ZANU-PF and the MDC remained.

Most blacks in Zimbabwe are farmers who raise only enough food for their families. Their main crop, corn, is pounded into flour to make corn bread.

GREAT ZIMBABWE

Since the late 1800's, Zimbabwe has suffered political troubles and violence, but the region has a long and great history that began thousands of years before the arrival of Europeans. In fact, the name Zimbabwe comes from an important part of that history.

Bushmen paintings and tools found in the region indicate that Stone Age people lived in what is now Zimbabwe. By the A.D. 800's, people were mining and trading minerals there. Then about A.D. 1000, the Shona people established their rule over the region.

The Shona built a city called Zimbabwe, or Great Zimbabwe, out of huge granite slabs that were skillfully fitted together, mostly without mortar. The word Zimbabwe means house of stone in chiShona, the language of the Shona.

Double walls skillfully constructed without the benefit of mortar may have provided a ceremonial walkway during the height of the empire.

The ruins of Great Zimbabwe lie near the modern town of Masvingo (once called Fort Victoria). They include a conical tower 30 feet (9 meters) high and part of a wall that rises up to 32 feet (10 meters) high and measures 800 feet (240 meters) around.

The city of Great Zimbabwe, the center of the Shona's Zimbabwe kingdom, eventually became the capital of two African empires. The first was the Mwanamutapa Empire, established during the 1400's by a branch of the Shona people called the Karanga. Their empire included what is now the country of Zimbabwe and part of Mozambique. At eastern African ports on the Indian Ocean, the Karanga traded ivory, gold, and copper for porcelain from China and for cloth and beads from India and Indonesia.

The Rozwi, a southern Karanga group, rebelled against the Mwanamutapa Empire in the late 1400's and conquered it. They founded their own empire, called the Changamire Empire, which became even stronger than the Mwanamutapa.

The Rozwi took over the city of Zimbabwe and built its largest structures. For more than 300 years, the Changamire Empire was prosperous and peaceful. Then in the 1830's, the empire was conquered by the Nguni people from the south, and the city of Great Zimbabwe was abandoned.

The conical tower still stands gracefully amid the ruins of Great Zimbabwe.

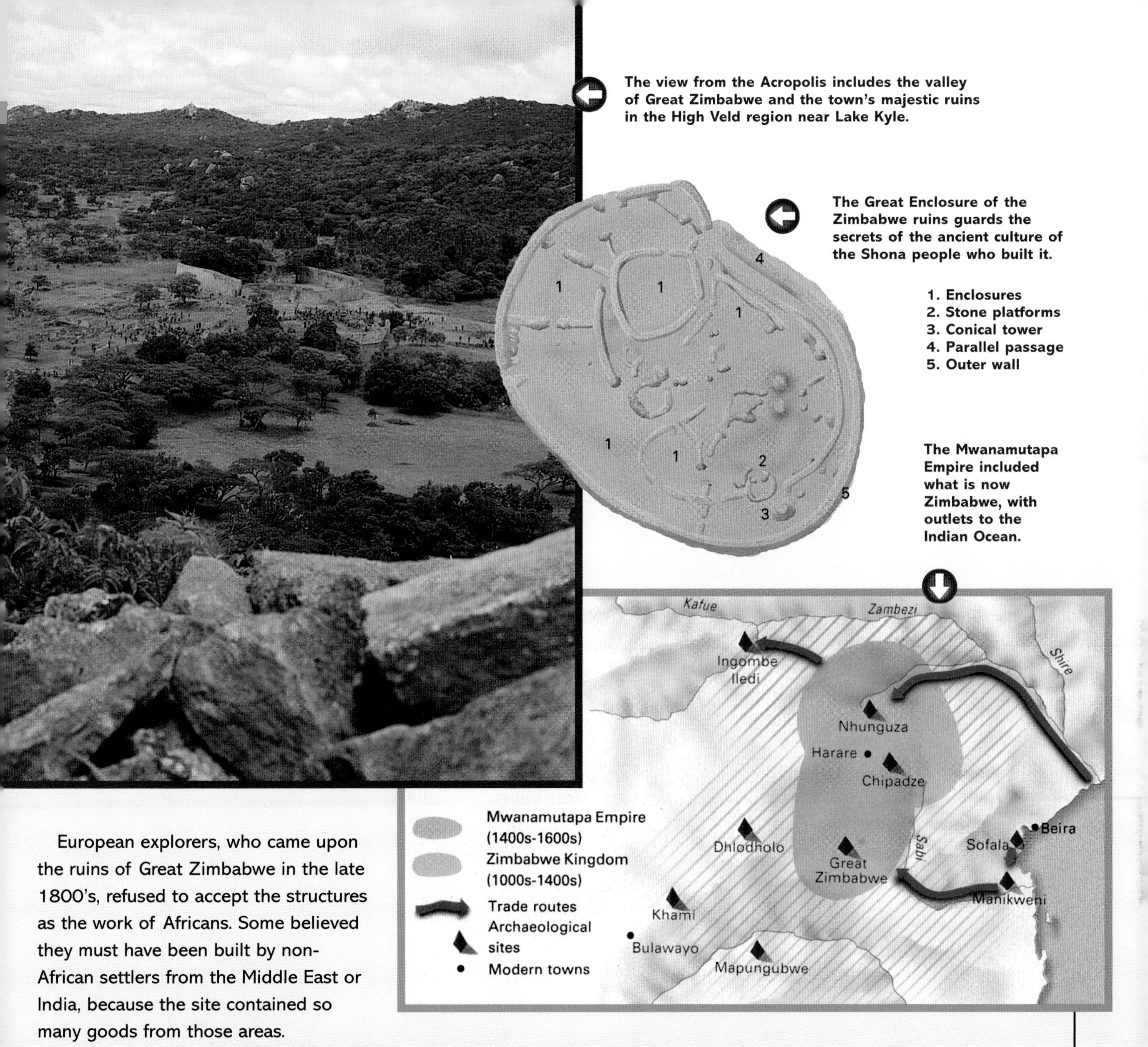

The view from the Acropolis includes the valley of Great Zimbabwe and the town's majestic ruins in the High Veld region near Lake Kyle.

The Great Enclosure of the Zimbabwe ruins guards the secrets of the ancient culture of the Shona people who built it.

The Mwanamutapa Empire included what is now Zimbabwe, with outlets to the Indian Ocean.

European explorers, who came upon the ruins of Great Zimbabwe in the late 1800's, refused to accept the structures as the work of Africans. Some believed they must have been built by non-African settlers from the Middle East or India, because the site contained so many goods from those areas.

However, no one had trouble believing that the ruins concealed vast quantities of gold. Treasure hunters plundered the ruins, stole the gold objects, and melted them down. In 1902, a law prevented the looters from ravaging the area any further. But so much had already been destroyed that there was little evidence left of the city's past.

Archaeologists have since determined that Great Zimbabwe thrived on foreign trade and that its buildings included the royal court, markets, warehouses, and religious shrines of the Shona people.

The British had named the whole region around the ruins Rhodesia, after the businessman Cecil Rhodes, whose company had gained control of the area. But when Rhodesia became a modern independent nation in 1980, its black majority chose the name Zimbabwe for their country.

Images of Great Zimbabwe are seen throughout the country today. A white triangle on the nation's flag contains a yellow Great Zimbabwe bird on a red star. Pictures of the conical tower appear on some of Zimbabwe's stamps and paper money.

Prospect Heights Public Library
12 N. Elm Street
Prospect Heights, IL 60070
www.phpl.info